Advances in Machine Learning and Data Analysis

Lecture Notes in Electrical Engineering

Volume 48

For other titles published in this series, go to
http://www.springer.com/series/7818

Sio-Iong Ao • Burghard B. Rieger
Mahyar Amouzegar

Editors

Advances in
Machine Learning
and Data Analysis

 Springer

Editors
Sio-Iong Ao
Harvard School of Engineering
and Applied Sciences
Harvard University
Room 403, 60 Oxford Street
Cambridge MA 02138, USA
siao@harvard.edu

Mahyar Amouzegar
College of Engineering
California State University
Long Beach

Burghard B. Rieger
Universität Trier
FB II Linguistische
Datenverarbeitung
Computerlinguistik
Universitätsring 15
54286 Trier
Germany
publication@iaeng.org

ISSN 1876-1100   e-ISSN 1876-1119
ISBN 978-94-007-3082-3   e-ISBN 978-90-481-3177-8
DOI 10.1007/978-90-481-3177-8
Springer Dordrecht Heidelberg London New York

Preface

A large international conference on Advances in Machine Learning and Data Analysis was held in UC Berkeley, CA, USA, October 22–24, 2008, under the auspices of the World Congress on Engineering and Computer Science (WCECS 2008). The WCECS is organized by the International Association of Engineers (IAENG). IAENG is a non-profit international association for the engineers and the computer scientists, which was founded in 1968 and has been undergoing rapid expansions in recent years. The WCECS conferences have served as excellent venues for the engineering community to meet with each other and to exchange ideas. Moreover, WCECS continues to strike a balance between theoretical and application development. The conference committees have been formed with over two hundred members who are mainly research center heads, deans, department heads (chairs), professors, and research scientists from over thirty countries. The conference participants are also truly international with a high level of representation from many countries. The responses for the congress have been excellent. In 2008, we received more than six hundred manuscripts, and after a thorough peer review process 56.71% of the papers were accepted.

This volume contains sixteen revised and extended research articles written by prominent researchers participating in the conference. Topics covered include Expert system, Intelligent decision making, Knowledge-based systems, Knowledge extraction, Data analysis tools, Computational biology, Optimization algorithms, Experiment designs, Complex system identification, Computational modeling, and industrial applications. The book offers the state of the art of tremendous advances in machine learning and data analysis and also serves as an excellent reference text for researchers and graduate students, working on machine learning and data analysis.

Harvard University, USA   Sio-Iong Ao
University of Trier, Germany   Burghard B. Rieger
California State University, Long Beach, USA   Mahyar Amouzegar

Contents

Chapter 1
2D/3D Image Data Analysis for Object Tracking and Classification

Seyed Eghbal Ghobadi, Omar Edmond Loepprich, Oliver Lottner,
Klaus Hartmann, Wolfgang Weihs, and Otmar Loffeld

Abstract Object tracking and classification is of utmost importance for different kinds of applications in computer vision. In this chapter, we analyze 2D/3D image data to address solutions to some aspects of object tracking and classification. We conclude our work with a real time hand based robot control with promising results in a real time application, even under challenging varying lighting conditions.

Keywords 2D/3D image data · Registration · Fusion · Feature extraction · Tracking · Classification · Hand-based robot control

1.1 Introduction

Object tracking and classification are the main tasks in different kinds of applications such as safety, surveillance, man–machine interaction, driving assistance system and traffic monitoring. In each of these applications, the aim is to detect and find the position of the desired object at each point in time. While in the safety application, the personnel as the desired objects should be tracked in the hazardous environments to keep them safe from the machinery, in the surveillance application they are tracked to analyze their motion behavior for conformity to a desired norm for social control and security. Man-Machine-Interaction, on the other hand has become an important topic for the robotic community. A powerful intuitive interaction between man and machine requires the robot to detect the presence of the user and interpret his gesture motion. A driving assistance system detects and tracks the obstacles, vehicles and pedestrians in order to avoid any collision in the moving path. The goal of traffic monitoring in an intelligent transportation system is to improve the efficiency and reliability of the transport system to make it safe and convenient

S.E. Ghobadi (✉), O.E. Loepprich, O. Lottner, K. Hartmann, W. Weihs, and O. Loffeld
Center for Sensor Systems (ZESS), University of Siegen, Paul-Bonatz-Str.9-11,
D57068 Siegen, Germany
e-mail: Ghobadi@zess.uni-siegen.de; Loepprich@zess.uni-siegen.de; Lottner@zess.uni-siegen.de;
Hartmann@zess.uni-siegen.de; Weihs@zess.uni-siegen.de; Loffeld@zess.uni-siegen.de

S.-I. Ao et al. (eds.), *Advances in Machine Learning and Data Analysis*,
Lecture Notes in Electrical Engineering 48, DOI 10.1007/978-90-481-3177-8_1,
© Springer Science+Business Media B.V. 2010

for the people. There are still so many significant applications in our daily life in which object tracking and classification plays an important role. Nowadays, the 3D vision systems based on Time of Flight (TOF) which deliver range information have the main advantage to observe the objects three-dimensionally and therefore they have become very attractive to be used in the aforementioned applications. However, the current TOF sensors have low lateral resolution which makes them inefficient for accurate processing tasks in the real world problems. In this work, we first propose a solution to this problem by introducing our novel monocular 2D/3D camera system and then we will study some aspects of object tracking and classification using 2D/3D image data.

1.2 2D/3D Vision System

Although the current optical TOF sensors [13–16] can provide intensity images in addition to the range data, they suffer from a low lateral resolution. This drawback can be obviated by combining a TOF camera with a conventional one. This combination is a tendency in the recent research works because even with regard to the emerging new generation of TOF sensors with high resolution,[1] an additional 2D sensor still results in a higher resolution and provides additional color information. With regard to the measurement range, however, the problem of parallax does not allow to simply position two cameras next to each other and overlay the generated images.

The multimodal data acquisition device used in this work is a recently developed monocular 2D/3D imaging system, named *MultiCam*. This camera, which is depicted in Fig. 1.1, consists of two imaging sensors: A conventional 10-bit CMOS gray scale sensor with VGA resolution and a Photonic Mixer Device (PMD) with a resolution of 64×48 pixels. The PMD is an implementation of an optical Time of Flight (TOF) sensor, able to deliver range data at quite high frame rates.

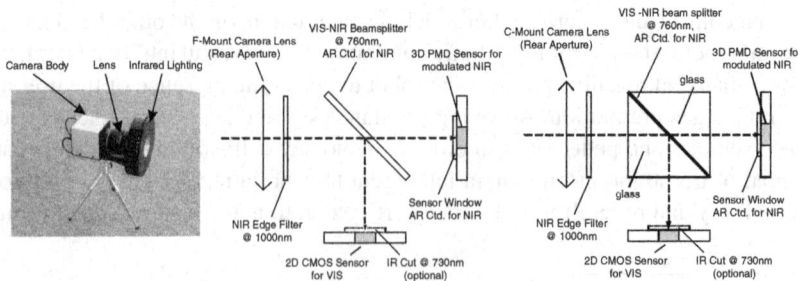

Fig. 1.1 *Left*: 2D/3D vision system (*MultiCam*) developed at ZESS. *Middle*: F-mount optical setup. *Right*: C-mount optical setup

---

[1] For example PMD-40K (200×200 pixels), Swissranger 4000 (176×144 pixels) and ZCam-prototype (320×480 pixels).

The principles of this sensor will be presented briefly in the next subsection. In addition, a beam splitter (see again Fig. 1.1), a near-infrared lighting system, a FPGA based processing unit as well as an USB 2.0 communication interface represent the remaining main components of this camera. It should be mentioned that the dichroic beam splitter behind the camera lens is used in order to divide the incident light into two spectral ranges: The visible part, which is forwarded to the CMOS chip and the near-infrared part to the TOF sensor [6]. Thus, the *MultiCam* is actually a multi-spectral device.

In fact, through the use of the *MultiCam*, one is able not just to achieve distance data at high frame rates (100 FPS and above) but also high resolution color images provided by the CMOS sensor. The novelty hereby is that a monocular setup is used which avoids parallax effects and makes the camera calibration a lot simpler along with the possibility to synchronize the 2D and 3D images down to several microseconds.

1.2.1 3D-Time of Flight Camera

Basically, the principle of the range measurement in a TOF camera relies upon the time difference Δt that the light needs to travel a distance d as follows

$$\Delta t = \frac{d}{c} \qquad (1.1)$$

where c represents the speed of light.

As a lighting source, we use a modulated light signal ($f_{mod} = 20\text{MHz}$), which is generated using a MOSFET based driver and a bank of high speed infrared emitting diodes. The illuminated scene then is observed by an intelligent pixel array (the PMD chip), where each pixel samples the amount of modulated light. To determine the distance d, we measure the phase delay $\Delta\varphi$ in each pixel. Recall that $\Delta\varphi = 2\pi \cdot f_{mod} \cdot \Delta t$ which in turn leads us to

$$d = \frac{c \cdot \Delta\varphi}{2\pi \cdot f_{mod}}. \qquad (1.2)$$

Since the maximal phase difference of $\Delta\varphi$ can be 2π, the unambiguous distance interval for range measurement at a modulation frequency of 20 MHz is equal to 15 m. This leads to the maximal (unambiguous) target distance of 7.5 m since the light has to travel the distance twice. In order to be able to use (1.2) for the distance computation in the TOF camera we have to multiply the equation by a factor of 0.5.

To calculate the phase delay $\Delta\varphi$, the autocorrelation function of the electrical an optical signal is analyzed by a phase-shift algorithm. Using four samples A_1, A_2, A_3 and A_4, each shifted by $\pi/2$, the phase delay can be calculated using [1]

$$\Delta\varphi = \arctan\left(\frac{A_1 - A_3}{A_2 - A_4}\right). \qquad (1.3)$$

In addition, the strength a of the signal, which in fact can be seen as its quality, along with the gray scale b can be formulated as follows [5]

$$a = \frac{1}{2} \cdot \sqrt{(A_1 - A_3)^2 + (A_2 - A_4)^2}, \qquad (1.4)$$

$$b = \frac{1}{4} \cdot \sum_{i=1}^{4} A_i. \qquad (1.5)$$

The environment lighting conditions in the background should be considered in all optical TOF sensors. There are various techniques dealing with this issue like using optical filters which only pass the band interested in, or applying some algorithm techniques that remove the noise artifacts of ambient light [8]. In our case, the PMD chip used has an in-pixel so-called SBI-circuitry (Suppression of Background Illumination) which increases the sensor dynamics under strong light conditions [1,13].

1.2.2 2D/3D Image Registration and Synchronization

As a prerequisite to profit from the 2D/3D multi-modality, the temporal and spatial relation of the individual sensors' images must be determined.

1.2.2.1 Temporal Synchronization

The detailed disquisition on the temporal synchronization of the individual sensors of the *MultiCam* points out that the camera's internal control unit (FPGA) can synchronize the 2D and the 3D sensor in the temporal domain within the limits of the clock resolution and minimal jitter due to the signal run times in the electronics. The synchronization can either refer to the beginning or to the end of the integration time of a 2D image and a single phase image. While the most common configuration is the acquisition of one 2D image per four phase images such that a new 2D image is available along with a new range image, it is also possible to acquire a new 2D image per phase image if necessary. Figure 1.2 gives an overview of the different possibilities.

If the synchronization of the 2D image relates to the first phase image, the temporal distance between the individual phase images is not equal, as the second phase image is captured only after the end of the 2D sensor's integration time. In contrast to that, synchronizing to the fourth phase image has the advantage of temporally equidistant phase images. In both configurations, it can occur that a change of the scene is represented only by one of both sensors if this change is outside of the actual integration time. With regard to the total time needed for the acquisition of a complete 2D/3D image, these two configurations do not differ from each other. However, the synchronization to the range image rather than to any of the phase

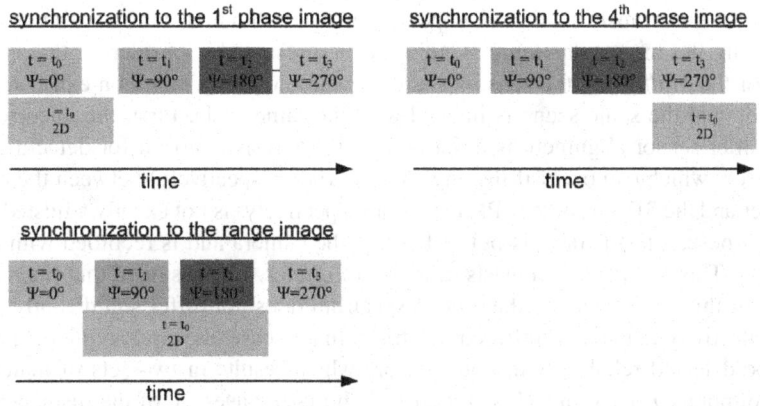

Fig. 1.2 Different possibilities of synchronizing of the 2D image to the 3D data

images is advantageous in that the total acquisition time is kept to a minimum, and in that the temporal equidistance of the phase images is maintained. The discussion on the motion artifacts in [10] gives details on the impacts of the individual configurations.

Binocular setups taking two complete cameras are evidently not as flexible and as precise as the very neatly controlled approach of the *MultiCam*.

1.2.2.2   Spatial Registration

In a 2D/3D vision system, regardless of the kind of setup, a 3D scene is imaged by two two-dimensional matrices of pixels with a degree of overlap which is a-priori unknown but qualitatively assumed to be high without loss of generality. Both sensors operate in a different spectrum (NIR vs. VIS) and have a different modality, i.e., the grey values of the scene represent different physical properties. Due to the operation principle, the sensors operate with different illumination sources meaning that the effects of the illumination must be taken into consideration (corresponding features may be dissimilar due to different lighting conditions). Both sensors have a different resolution with the 2D sensor's resolution being higher. The TOF sensor acquires the distance to an observed point with an accuracy and reproducibility in the range of a centimeter. The relative arrangement of both sensors is not a function of the time but is known in advance with only an insufficient accuracy, meaning that the configuration needs to be calibrated initially.

The aim of the image registration is to establish a spatial transform that maps points from one image to homologous points in a target image as follows

$$[x_1, y_1, z_1]^T = f\left([x_2, y_2, z_2]^T\right). \tag{1.6}$$

The actual transform model depends on the individual configuration. In the following the monocular setup is going to be presented. Considering the special case of the *MultiCam*, the sensors share a common lens, a common extrinsic calibration and the same scene is imaged with the same scale. First, the uncorrected view after sensor alignment is described. This analysis is useful for detecting angle errors which can occur if the angle of 45° (90° respectively) between the beam splitter and the 2D sensor (the PMD sensor respectively) is not exactly adjusted. For this purpose, a test pattern is put in front of the camera and is recorded with both sensors. This test pattern consists of a grid of circles. It is assumed that the reflectivity of this test pattern in the visible spectrum does not differ significantly from the reflectivity in the near-infrared spectrum. In that case, the circles' middle points can be detected reliably with both sensors which results in two sets of matching control points P_{PMD} and P_{2D}. Figure 1.3 shows the average of the displacement between these two sets in units of 2D pixels as a function of the distance between the camera and the pattern for a constant focal length. The average and the standard deviation are computed out of all the circles' middle points in the image. It can be observed that the displacement averages are stable over distance, which means that there is virtually no angle error in the sensor alignment in the observed range. By examining the displacement distribution over the whole image, it can be further concluded that the displacement arrangement does not reveal significant local deviations. Consequently a global rigid transformation model can be used which is independent of the location in the image. Additionally, the uncorrected view shows that the pixel-to-pixel mapping is fixed with a negligible rotational component and which, in particular, is independent from the depth. What remains is a transformation composed only of a two-dimensional translational displacement. Consequently, an iterative closest point algorithm[2] is used to find an optimal solution.

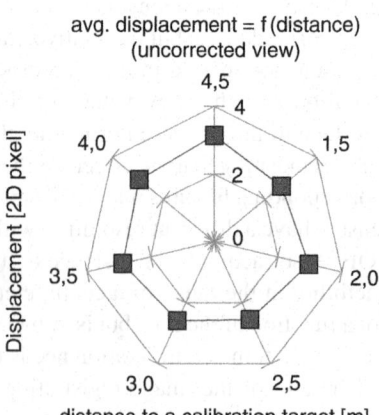

Fig. 1.3 Uncorrected dislocation of PMD and 2D sensor in the *MultiCam*

avg. displacement = f (distance)
(uncorrected view)

Displacement [2D pixel]

distance to a calibration target [m]

---

[2] Levenberg-Marquardt algorithm; the optimization criterion is the sum of squared distances of the individual points.

1.3   Multimodal Data Fusion and Segmentation

The TOF camera delivers three data items for each pixel at each time step: intensity, range and amplitude of the received modulated light. The intensity image of the TOF camera comparable to the intensity images in CCD or CMOS cameras relies on the environment lighting conditions, whereas the range image and the amplitude of the received modulated light are mutually dependent.

None of these individual data can be used solely to make a robust segmentation under variant lighting conditions. Fusing these data provides a new feature information which is used to improve the performance of the segmentation technique.

In this paper we have used the basic technique for the fusing of the range and intensity data which has already been used in other fields like SAR imaging. We observed that the range data in our TOF sensor is dependent on the reflection factor of the object surface (how much light is reflected back from the object). Therefore, there is a correlation between the intensity and range vector sets in a TOF image. These two vector sets are fused to derive a new data set, so-called "phase", which indicates the angle between two intensity and range vector sets. The details of this technique is presented in our previous works [12]. Another type of fusion which has also been used in our work is to weight the value of the range for each pixel using the modulation amplitude which adjusts the range level in the regions where the range data might get wrong.

However, using *MultiCam*, we can acquire low resolution TOF images with their corresponding features derived from fusion; and high resolution 2D Images. For segmentation, same as in [12], first we apply the unsupervised clustering technique to segment the low resolution TOF images. Next, we map the 3D segmented image to 2D image. Due to the monocular setup of *MultiCam*, mapping the 3D range image to the 2D image is a trivial and fast task which consequently makes the segmentation of high resolution 2D image computationally cheap. This kind of segmentation has two main advantages over 2D segmentation. On the one hand 3D range segmentation is more reliable and robust in the natural environment where lighting conditions might change and on the other hand due to the low resolution of 3D image, segmentation is faster. An example of such a segmentation is shown in Fig. 1.4.

1.4   Object Tracking and Classification

One of the approach for object identification in tracking process is to use a classifier directly to distinguish between different detected objects. In fact, if the classification method is fast enough to operate at image acquisition frame rate, it can be used directly for tracking as well. For example, supervised learning techniques such as Support Vector Machines (SVM) and AdaBoost can be directly employed to classify the objects in each frame because they are fast techniques which can work at real time rate for many applications.

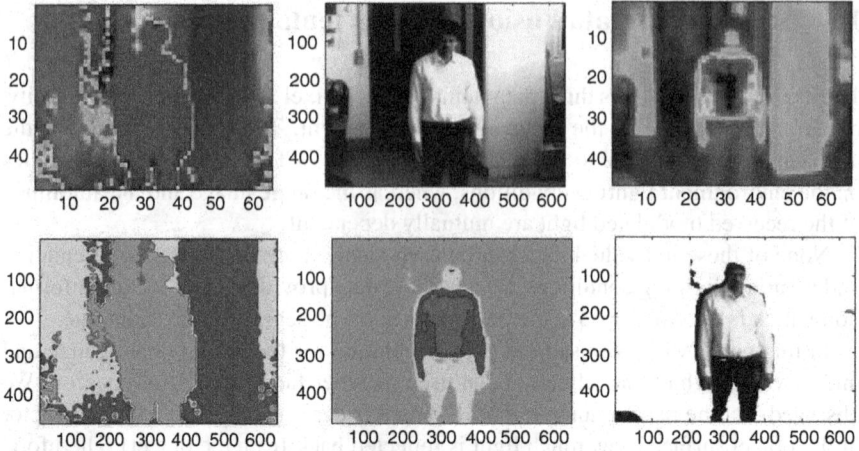

Fig. 1.4 Segmentation in 2D/3D images. *Top Left*: low resolution range image from TOF sensor. *Top Middle*: high resolution 2D image. *Top Right*: modulation amplitude image. *Bottom Left*: 3D rescaled segmented image. *Bottom Middle*: rescaled segmented image using fusion of range and modulation amplitude data. *Bottom Right*: 2D segmented image result from mapping

In this section, we describe tracking with classifier more in detail by applying a supervised classifier based on AdaBoost to 2D/3D videos in order to detect and track the desired object. After segmentation of the image which was described in the previous section, in the next step the Haar-Like features are extracted and used as the input data for the AdaBoost classifier. Haar-like features which have been used successfully in face tracking and classification problems encode some information about the object to be detected. For a much more in depth understanding the reader is referred to [11].

However, there are two main issues in real time object detection based on Haar-Like features and using AdaBoost technique. The first issue is that background noise in the training images degrades detection accuracy significantly, esp. when it is a cluttered background with varying lighting condition which is the case in many real world problems. The second issue is that computation of all sub-windows (search windows) in an image for every scale is too costly if the real time constraints are to be met. The fundamental idea of our algorithm is to address the solution to these problems using fusion of 3D range data with 2D images. In order to extinguish the background issue from object recognition problem, the procedure of object detection is divided into two levels. In the low level we use range data in order to: (i) Define a 3D volume where the object of interest is appearing (Volume of Interest) and eliminate the background to achieve robustness against cluttered backgrounds and (ii) Segment the foreground image into different clusters. In the high level we map the 3D segmented image to its corresponding 2D color image and apply Viola-Jones method [11] (searching with Haar-Like features) to find the desired object in the image. Figure 1.5 shows some examples of this procedure for hand detection and tracking.

Original 2D Image   3D Range Image   3D Segmented Image   2D Segmented Image

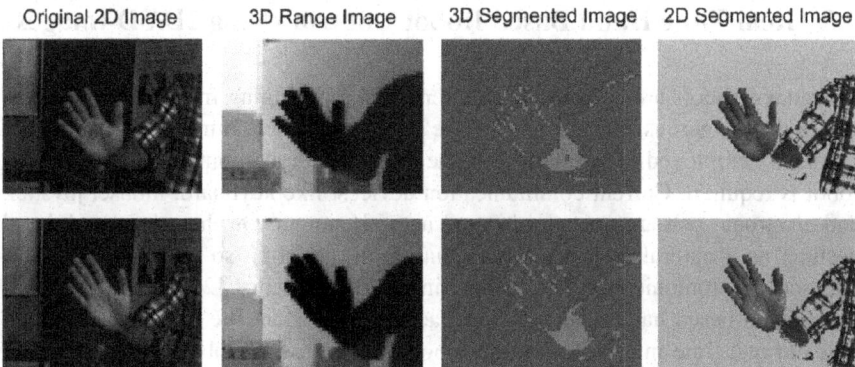

Fig. 1.5 Solution to the background issue in object detection using Viola-Jones method. Using range data the cluttered background is removed and the foreground image is segmented and mapped to 2D image. Viola-Jones technique is applied to 2D segmented image to find the object of interest

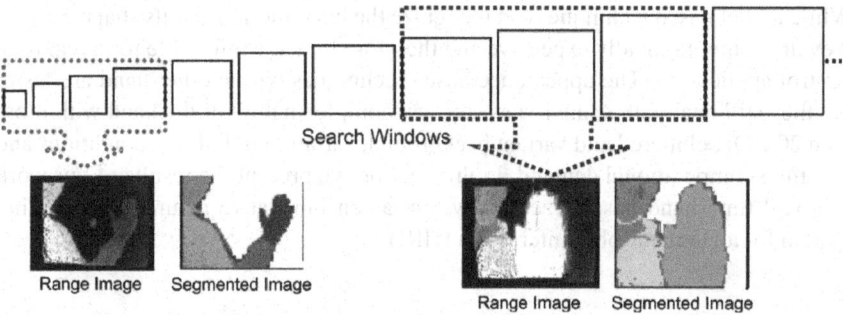

Fig. 1.6 Selection of search windows using range information for hand detection. *Left*: Hand is far from the camera and therefore the image is searched with small search windows. *Right*: Hand is close to the camera and therefore the image is scanned with large search windows to find the hand in the image

The second issue (Computation of all search windows in an image for every scale is too costly.) can be addressed by using the range information directly. After segmentation, the distance of the segmented object from the camera can be easily derived from 3D range image. By having the information about the distance of object from the camera, its size can be roughly estimated and a set of search windows which could fit to the size of the object is selected and therefore there is no need to use all possible size of search windows to find the object. This reduces the computational cost of the Viola-Jones technique to a great extent which is a significant point in real time applications. An example of selecting search windows for hand detection is illustrated in Fig. 1.6.

1.5 Real Time Hand Based Robot Control Using 2D/3D Images

Nowadays, robots are used in the different domains ranging from search and rescue in the dangerous environments to the interactive entertainments. The more the robots are employed in our daily life, the more a natural communication with the robot is required. Current communication devices, like keyboard, mouse, joystick and electronic pen are not intuitive and natural enough. On the other hand, hand gesture, as a natural interface means, has been attracting so much attention for interactive communication with robots in the recent years [2–4, 9]. In this context, vision based hand detection and tracking techniques are used to provide an efficient real time interface with the robot. However, the problem of visual hand recognition and tracking is quite challenging. Many early approaches used position markers or colored gloves to make the problem of hand recognition easier, but due to their inconvenience, they can not be considered as a natural interface for the robot control. Thanks to the latest advances in the computer vision field, the recent vision based approaches do not need any extra hardware except a camera. These techniques can be categorized as: model based and appearance based methods [7]. While model based techniques can recognize the hand motion and its shape exactly, they are computationally expensive and therefore they are infeasible for a real time control application. The appearance based techniques on the other hand are faster but they still deal with some issues such as: complex nature of the hand with more than 20 DOF, cluttered and variant background, variation in lighting conditions and real time computational demand. In this section we present the results of our work in a real time hand based tracking system as an innovative natural commanding system for a Human Robot Interaction (HRI).

1.5.1 Set-Up

Set-up mainly consists of three parts: (1) A six axis, harmonic driven robot from *Kuka* of type KR 3 with attached magnetic grabber. The robot itself has been mounted onto an aluminium rack along with the second system component. (2) A dedicated robot control unit, responsible for robot operation and communication by running proprietary software from $Kuka^{©}$ company. (3) The main PC responsible for data acquisition from 2D/3D imaging system (*MultiCam*) and running the algorithms. Communication between the robot control unit and the application PC is done by exchanging XML-wrapped messages via TCP/IP. The network architecture follows a strict client server model, with the control unit as the client connecting to the main PC, running a server thread, during startup.

1.5.2 Control Application

In order to make the communication system more convenient for the user, all the necessary commands to control the robot, such as moving the robot in 6 directions $(x^+, x^-, y^+, y^-, z^+, z^-)$ or (de)activating the grabber (palm-to-fist or vice versa) are done by using a self developed GUI based application illustrated in Fig. 1.7. As a first step, we track the user's hand movement in a predefined volume covered by the *MultiCam*, followed by mapping its real world position into a virtual space which is represented by a cuboid of defined size and correlates with the *MultiCam's* view frustum. Hand movement is visualized by placing a 3D hand-model in the according location within the cuboid. Depending on the hand's distance from the cuboid's center, a velocity vector is generated, along with some other state information, and sent to the robot's control unit which is in charge of sending the appropriate information to the robot itself. By placing the virtual hand in the cuboid's center, the system can be put in a susceptible mode for special commands. For that matter, a rudimentary gesture classification algorithm has been implemented which is able to distinguish between a fist and a palm. We use defined fist to palm transition sequences (e.g., a palm-fist-palm transition) in order to perform a robot reset, put the system in predefined modes and to (de)activate the magnetic grabber which in turn enables the whole system to handle ferric objects (Fig. 1.8).

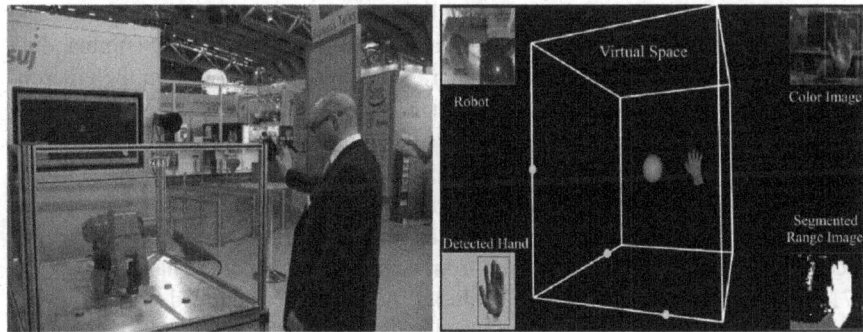

Fig. 1.7 *Left*: Hand based robot control using *MultiCam*, Hannover Fair 2008, *Right*: Graphical User Interface (GUI)

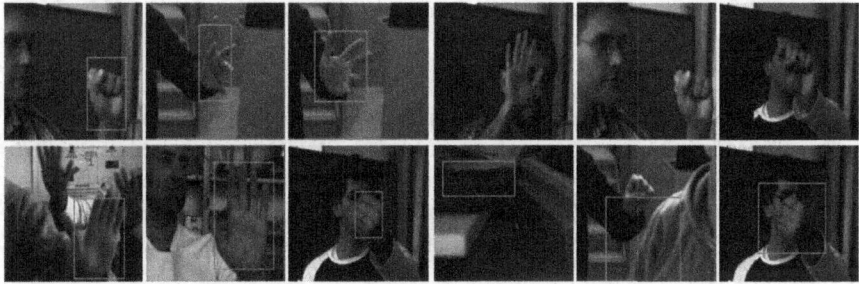

Fig. 1.8 Some results, *left*: example of correctly detected images (True Positive), *right*: example of wrongly detected images (*First row*: missed hand, *Second row*: misclassified)

Table 1.1 Confusion table
for hand detection system

	Hand	Non-hand
Hand	2,633	87
Non-hand	224	2,630
Sum	2,857	2,717

1.5.3 Experimental Results

For the Hannover Fair 2008, a simple task had been defined to be performed by the visitors and to put the system's performance under the test as follows: Commanding the robot to move in six directions using moving the hand with any kind of posture in the corresponding directions, picking up a metal object with the magnet grabber using palm to fist gesture, moving the object using the motion of the hand and finally dropping it in the defined areas with palm to fist gesture. It turned out that the system handling has been quite intuitive, since different people have been able to operate the robot instantly. In terms of reliability the whole system worked flawlessly during the complete time exposed at the fair. For training of the classifier we took 1037 positive hand images from seven people, and 1,269 negative images from non-hand objects in our lab environment. Using OpenCV we trained our classifier with 20 stages and window size of 32×32. Although the classifier was trained under the lab conditions, it worked quite well under the extreme lighting conditions at the fair.

In order to analyze the performance of the system, we recorded the results of hand detection from our GUI in the video format while different users were commanding the robot. Likewise, we moved the camera and took the videos from the environment. These videos are labeled as "Positive" and "Negative" data. While positive stands for the hand, the negative represents the non-hand objects. The data were acquired using a PC with dual core 2.4 GHz CPU. The exposure time for 3D sensor was set at 2ms while for 2D sensor it was about 10 ms. Under these conditions, we had about 15 detected images (including all algorithms computational time) per second. The confusion matrix derived from these videos with 2857 hand images and 2717 non-hand images is shown in Table 1.1. As it can be calculated from this table, the system has a Hit Rate of 0.921, False Positive Rate of 0.032 and the recognition accuracy of 94.4%.

1.6 Conclusion

In this work we study some aspects of object detection and tracking using 2D/3D Images. These images are provided by a monocular 2D/3D vision system, so-called *MultiCam*. The principle of this camera system as well as the registration and fusion of 2D/3D data are discussed. This work is concluded with some results of a real time hand based robot control application which was demonstrated at Hannover fair in Germany in 2008.

Acknowledgments This work has been funded by German Research Foundation (DFG) under contract number LO 455/10-2 which is gratefully appreciated.

References

1. Moeller, T. Kraft, H. and Frey, J.: Robust 3D Measurement with PMD Sensors, PMD Technologies GmbH, www.pmdtec.com
2. Wang, C.C. and Wang, K.C.: Hand Posture Recognition Using Adaboost with SIFT for Human Robot Interaction, International Conference on Advanced Robotics, 2007.
3. Cerlinca, T.L., Pentiuc, S.G. and Cerlinca, M.C.: Hand Posture Recognition for Human-Robot Interaction. Proceedings of the 2007 workshop on Multimodal interfaces in semantic interaction,2007.
4. Malima, A., Ozgur, E. and Cetin, M.: A Fast Algorithm for Vision-based Hand Gesture Recognition for Robot Control. IEEE Conference on Signal Processing and Communications Applications, 2006.
5. Ghobadi, S.E., Hartmann, K., Weihs, W., Netramai, C., Loffeld, O. and Roth, H.: Detection and Classification of Moving Objects-Stereo or Time-of-Flight Images, IEEE conference on Computational Intelligence and Security, 2006, China.
6. Lottner, O., Hartmann, K., Weihs, W. and Loffeld, O.: Image Registration and Calibration aspects for a new 2D/3D camera, EOS Conference on Frontiers in Electronic Imaging, June 2007, Munich, Germany.
7. Fang, Y., Wang, K., Cheng, J. and Lu, H.: A Real-Time Hand Gesture Recognition Method. 2007 IEEE International Conference on Multimedia and Expo.
8. Gokturk, S.B., Yalcin, H. and Bamji, C.: A Time of Flight Depth Sensor, System Description, Issues and Solutions, on IEEE workshop on Real-Time 3D Sensors and Their Use in conjunction with IEEE Conference on Computer Vision and Pattern Recognition, CVPR, Washington, USA 2004.
9. Rogalla, O., Ehrenmann, M., Zoellner, R., Becher, R. and Dillmann, R.: Using Gesture and Speech Control for Commanding a Robot Assistant. 11th IEEE International Workshop on Robot and Human Interactive Communication, 2002.
10. Lottner, O., Sluiter, A., Hartmann, K. and Weihs, W.: Movement Artefacts in Range Images of Time-of-Flight Cameras, EOS DOI: 10.1109/ISSCS.2007.4292665, 2007 Romania.
11. Viola, P. and Jones, M.: Rapid Object Detection using a Boosted Cascade of Simple Features. Conference on Computer vision and Pattern Recognition, 2001.
12. Ghobadi, S.E., Loepprich, O., Hartmann, K. and Loffeld, O.: Hand Segmentation Using 2D/3D Images, IVCNZ 2007 Conference, Hamilton, New Zealand, 5–7. December, 2007.
13. PMD-Technologie; www.pmdtec.com
14. Swissranger; C.C.S. d'Electronique SA, http://www.mesa-imaging.ch
15. Canesta, Inc., http://www.canesta.com/
16. 3DV Systems, ZCam; http://www.3dvsystems.com/

Chapter 2
Robot Competence Development by Constructive Learning

Q. Meng, M.H. Lee, and C.J. Hinde

Abstract This paper presents a constructive learning approach for developing sensor-motor mapping in autonomous systems. The system's adaptation to environment changes is discussed and three methods are proposed to deal with long term and short term changes. The proposed constructive learning allows autonomous systems to develop network topology and adjust network parameters. The approach is supported by findings from psychology and neuroscience especially during infants cognitive development at early stages. A growing radial basis function network is introduced as a computational substrate for sensory-motor mapping learning. Experiments are conducted on a robot eye/hand coordination testbed and results show the incremental development of sensory-motor mapping and its adaptation to changes such as in tool-use.

Keywords Developmental robotics · Biologically inspired systems · Constructive learning · Adaptation

2.1 Introduction

In many situations such as home services for elderly and disabled people, artificial autonomous systems (e.g., robots) need to work for various tasks in an unstructured environment, system designers cannot anticipate every situation and program the system to cope with them. This is different from the traditional industrial robots which mostly work in structured environments and are programmed each time for a specific task. Autonomy, self-learning and organizing, and adapting toenvironment changes are crucial for these artificial systems to successfully

Q. Meng (✉) and C.J. Hinde
Department of Computer Science, Loughborough University, LE11 3TU, UK
e-mail: q.meng@lboro.ac.uk

M.H. Lee
Department of Computer Science, University of Wales, Aberystwyth, SY23 3DB, UK

S.-I. Ao et al. (eds.), *Advances in Machine Learning and Data Analysis*,
Lecture Notes in Electrical Engineering 48, DOI 10.1007/978-90-481-3177-8_2,
© Springer Science+Business Media B.V. 2010

fulfil various challenging tasks. Traditional controllers for intelligent systems are designed by hand, and they do not have such flexibility and adaptivity. General cognitivist approach for cognition is based on symbolic information processing and representation, and does not need to be embodied and physically interact with the environment. Most cognitivist-based artificial cognitive systems rely on the experience from human designers.

Human beings [1] and animals face similar problems during their development of sensor-motor coordination, however we can tackle these problems without too much effort. During human cognitive development, especially at the early stages, each individual undergoes changes both physically and mentally through interaction with environments. These cognitive developments are usually staged, exhibited as behavioural changes and supported by neuron growth and shrinking in the brain. Two kinds of developments in the brain support the sensory-motor coordination: quantitative adjustments and qualitative growth [19]. Quantitative adjustments refer to the adjustments of the synapse connection weights in the network and qualitative growth refers to the changes of the topology of the network. Inspired by developmental psychology especially Piaget's sensory-motor development theory of infants [12], developmental robotics focuses on mechanisms, algorithms and architectures for robots to incrementally and automatically build their skills through interaction with their environment [21]. The key features of developmental robotics share similar mechanisms with human cognitive development which include learning through sensory-motor interaction; scaffolding by constraints; staged, incremental and self-organizing learning; intrinsic motivation driven exploration and active learning; neural plasticity, task transfer and adaptation. In this paper, we examine robot sensory-motor coordination development process at early stages through a constructive learning algorithm. Constructive learning which is inspired by psychological constructivism, allows both quantitative adjustments and qualitative network growth to support the developmental learning process. Most static neural networks need to predefine the network structure and learning can only affect the connection weights, and they are not consistent with developmental psychology. Constructive learning is supported by recent neuroscience findings of synaptogenesis and neurogenesis occurring under pressures to learn [16, 20]. In this paper, a self-growing radial basis function network (RBF) is introduced as the computational substrate, and a constructive learning algorithm is utilized to build the sensory-motor coordination development. We investigate the plasticity of the network in terms of self-growing in network topology (growing and shrinking) and adjustments of the parameters of each neuron: neuron position, the size of receptive field of each neuron, and connection weights. The networks adaptation to systems changes is further investigated and demonstrated by eye/hand coordination test scenario in tool-use.

2.2   Sensory-Motor Mapping Development Via Constructive Learning

In order to support the development of sensor-motor coordination, a self-growing RBF network is introduced due to its biological plausibility. There exists very strong evidence that humans use basis functions to perform sensorimotor transformations [15], Poggio proposed that the brain uses modules as basis components for several of its information processing subsystems and these modules can be realized by generalized RBF networks [13, 14].

There are three layers in the RBF network: input layer, hidden layer and output layer. The hidden layer consists of radial basis function units (neurons), the size of receptive field of each neuron varies and the overlaps between fields are different. Each neuron has its own centre and coverage. The output is the linear combination of the hidden neurons.

A RBF network is expressed as:

$$\mathbf{f}(\mathbf{x}) = \mathbf{a}_0 + \sum_{k=1}^{N} \mathbf{a}_k \phi_k(\mathbf{x}) \tag{2.1}$$

$$\phi_k(\mathbf{x}) = \exp\left(-\frac{1}{\sigma_k^2} \|\mathbf{x} - \mu_k\|^2\right) \tag{2.2}$$

where $\mathbf{f}(\mathbf{x}) = (f_1(\mathbf{x}), f_2(\mathbf{x}), \cdots, f_{N_o}(\mathbf{x}))^T$ is the vector of system outputs, N_o is the number of outputs and $\mathbf{X}$ is the system input. $\mathbf{a}_k$ is the weight vector from the hidden unit $\phi_k(\mathbf{x})$ to the output, N is the number of radial basis function units, and μ_k and σ_k are the kth hidden unit's center and width, respectively.

2.2.1   Why Constructive Learning?

According to Shultz [19, 20], in addition to that constructive learning is supported by biological and psychological findings, there are several advantages of constructive learning over static learning: first, constructive-network algorithms learn fast (in polynomial time) compared with static learning (exponential time), and static learning maybe never solve some problems as the designer of a static network must first find a suitable network topology. Second, constructive learning may find optimal solutions to the bias/variance tradeoff by reducing bias via incrementally adding hidden units to expand the network and the hypothesis space, and by reducing variance via adjusting connection weights to approach the correct hypothesis. Third, static learning cannot learn a particular hypothesis if it has not been correctly represented, a network may be too weak to learn or too powerful to generalize. Constructive learning avoids this problem because its network growth enables it to represent a hypothesis that could not be represented previously with limited network power.

2.2.2 Topological Development of the Sensory-Motor Mapping Network

During the development of sensory-motor mapping network, two mechanisms exist: topological changes of the mapping network and network parameter adjustments. The qualitative growth of the sensory-motor mapping network depends on the novelty of the sensory-motor information which the system obtained during its interaction with the environment in development, the growth is incremental and self-organizing. The sensory-motor mapping network starts with no hidden units, and with each development step, i.e., after the system observes the consequence of an action, the network grows or shrinks when necessary or adjusts the network parameters accordingly. The network growth criteria are based on the novelty of the observations, which are: whether the current network prediction error for the current learning observation is bigger than a threshold, and whether the node to be added is far enough from the existing nodes in the network: $\|\mathbf{e}(t)\| = \|\mathbf{y}(t) - \mathbf{f}(\mathbf{x}(t))\| > e_1$, $\|\mathbf{x}(t) - \boldsymbol{\mu}_r(t)\| > e_3$. In order to ensure smooth growth of the network the prediction error is checked within a sliding window: $\sqrt{\sum\limits_{j=t-(m-1)}^{t} \frac{\|\mathbf{e}(j)\|^2}{m}} > e_2$, where, $(\mathbf{x}(t), \mathbf{y}(t))$ is the learning data at tth step, and $\boldsymbol{\mu}_r(t)$ is the centre vector of the nearest node to the current input $\mathbf{x}(t)$. m is the length of the observation window. If the above three conditions are met, then a new node is inserted into the network with the following parameters: $\mathbf{a}_{N+1} = \mathbf{e}(t)$, $\boldsymbol{\mu}_{N+1} = \mathbf{x}(t)$, $\sigma_{N+1} = k \|\mathbf{x}(t) - \boldsymbol{\mu}_r(t)\|$, where, k is the overlap factor between hidden units.

The above network growth strategy does not include any network pruning, which means the network size will become large, some of the hidden nodes may not contribute much to the outputs and the network may become overfit. In order to overcome this problem, we use a pruning strategy as in [8], over a period of learning steps, to remove those hidden units with insignificant contribution to the network outputs.

Let o_{nj} be the jth output component of the nth hidden neuron, $o_{nj} = a_{nj} \exp(-\frac{\|\mathbf{x}(t) - \boldsymbol{\mu}_n\|^2}{\sigma_n^2})$, $r_{nj} = \frac{o_{nj}}{max(o_{1j}, o_{2j}, \cdots, o_{Nj})}$.

If $r_{nj} < \delta$ for M consecutive learning steps, then the nth node is removed. δ is a threshold.

2.2.3 Parameter Adjustments of the Sensory-Motor Mapping Network

There are two types of parameters in the network, the first type of parameter is the connection weights; the second is parameters of each neuron in the network: the position and the size of receptive field of each neuron. A simplified node-decoupled EKF (ND-EKF) algorithm was proposed to update the parameters of each node independently in order to speed up the process. The parameters of the network are grouped into $N_o + N$ components. The first N_o groups are the weights, $\mathbf{w}_k = [a_{0k}, a_{1k}, \cdots, a_{Nk}]^T, k = 1, 2, \cdots, N_o$ (a_{ij} is the weight from ith hidden node to jth output); and the rest N groups are the parameters of hidden units' parameters: $\mathbf{w}_k = [\boldsymbol{\mu}_k^T, \sigma_k]^T, k = 1, 2, \cdots, N$. The superscript T stands for transpose of a matrix.

So for kth parameter group at tth learning step, ND-EKF is given by:

$$\mathbf{w}_k(t) = \mathbf{w}_k(t-1) + \mathbf{K}_k(t)\mathbf{e}_k(t) \qquad (2.3)$$

where

$$\mathbf{e}_k(t) = \begin{cases} y_k(t) - f_k(\mathbf{x}(t)) & k = 0, 1, 2, \cdots, N_o \\ \mathbf{y}(t) - \mathbf{f}(\mathbf{x}(t)) & k = N_o + 1, \cdots, N_o + N \end{cases} \qquad (2.4)$$

and $\mathbf{K}_k(t)$ is the kalman gain, $y_k(t)$ is the kth component of $\mathbf{y}(t)$ in training data $(\mathbf{x}(t), \mathbf{y}(t))$, $\mathbf{B}_k(t)$ is the submatrix of derivatives of network outputs with respect to the kth group's parameters at tth learning step. $\mathbf{R}_k(t)$ is the variance of the measurement noise, and is set to be $diag(\lambda)$ (λ is a constant) in this paper. q is a scalar that determines the allowed random step in the direction of the gradient vector.

In our algorithm, an extended Kalman filter is used to adjust the systems's parameters. There may exist a similar mechanism in our brain. Recent research findings has found evidences that Kalman filtering occurs in visual information processing [17, 18], motor coordination control [22], and spatial learning and localization in the hippocampus [1, 21]. In hippocampus studies, a Kalman filtering framework has been mapped to the entorhinal-hippocampal loop in a biologically plausible way [1, 21]. According to the mapping, region CA1 in the hippocampus holds the system reconstruction error signal, and the internal representation is maintained by Entorhinal Cortex (EC) V–VI. The output of CA1 corrects the internal representation, which in turn corrects the reconstruction of the input at EC layers II–III. O'Keefe also provided a biologically plausible mechanism by which matrix inversions might be performed by the CA1 layer through an iterated update scheme and in conjunction with the subiculum [11]. In addition, the matrix inversion lemma has been widely used in computational neuroscience [4].

2.3   Adaptation of Sensory-Motor Mapping

Two kinds of changes in our daily life may require the learned sensory-motor mapping to update: short term changes and long term changes. For the short term, humans may just reuse learned knowledge and quickly adjust some parameters to adapt to the environment changes. But for the longer term, after an adult is trained in a special environment or for a special tasks for a long time, they may grow new neurons to gain new skills, and to enhance the already acquired knowledge. Examples of these two kinds of changes can be found during human development, the kinematics of limbs and bodily structures are not fixed during human growth but may change, either slowly over long periods during growth and bodily maturation, or rapidly such as when we use tools to extend the reach or function of our manipulation abilities. It has been discovered that infants learn and update sensorimotor mappings by associating spontaneous motor actions and their sensory consequences [12]. It takes a relatively long time to build up the mapping skills, which involves neuron growth processes in the brain to support the sensorimotor transformation. After an adult has gained the basic skills, they can quickly adapt to different situations, for example, an adult can quickly adapt to the use of a pointer to point to a seen target. This indicates that after rapid structural changes we do not learn new sensorimotor skills from scratch, rather we reuse the existing knowledge and simply (and quickly) adjust some parameters. Maguire et al [9] studied the structural changes in the hippocampi of licensed London tax drivers. They found that taxi drivers had a significantly greater volume in the posterior hippocampus, whereas control subjects showed greater volume in the anterior hippocampus. Maguire's study suggests that the human brain grows or shrinks to reflect the cognitive demands of the environment, even for adults.

In autonomous systems, some parameters may gradually change after a long time use, the systems need to adapt to these changes automatically. Autonomous systems have additional situations where structures may change suddenly, these may be unintentional, for example when damage occurs through collisions, or by design when a new tool is fitted to the arm end-effector. For these reasons it is important for autonomous systems in unstructured environments to have the ability to quickly adjust the existing mapping network parameters so as to automatically re-gain the eye/hand coordination skills. We note that humans can handle this problem very well. Recent neurophysiological, psychological and neuropsychological research provides strong evidence that temporal, parietal and frontal areas within the left cerebral hemisphere in humans and animals are involved and change during activities where the hand has been extended physically, such as when using tools [2, 3, 5, 6, 10]. Japanese macaque monkeys were trained to use a rake to pull food closer, which was originally placed beyond the reach of their hands [2, 3]. The researchers found that, in monkeys trained in tool-use, a group of bimodal neurons in the anterior bank of the intraparietal sulcus, which respond both to somatosensory and visual stimuli related to the hand, dynamically altered their visual receptive field properties (the region where a neuron responds to certain visual stimuli) during training of the tool-use.

In this paper, we develop approaches of adapting to environments, and more specifically, different robot limb sizes in our experiments, were investigated and compared. All these adaptation skills are usually not available in commercial calibration-based eye/hand mapping systems.

In our plastic RBF network for robotic eye/hand mapping, the knowledge learned for the mapping is stored in the network in terms of the number of neurons, their positions and sizes of receptive fields, and the node weights. In order to quickly adapt to structural changes of the robotic system, this knowledge needs to be reused in some way rather than setting up the network again from empty. In this paper, we considered three methods for such adaptation, all of them reuse the learned knowledge by adjusting the learned network:

1. Full adjustment of the learned network after a structural change. This includes network topological changes by adding new hidden nodes or remove existing

ones if necessary, and adjusting the following parameters: the centres and widths of the existing nodes, and the weights from the hidden nodes to the outputs.

2. Adjusting the weights of the learn+. ed network, removing the insignificant hidden units, but keeping the rest of the hidden units unchanged.
3. Only adjusting the weights, and keeping the hidden unit structure of the learned network completely unchanged.

2.4 Experimental Studies

2.4.1 Experimental System

In this paper, the robot eye/hand coordination is used as a testbed to demonstrate the process of constructive learning and adaptation of the sensory-motor mapping network to the changes. The experimental robot system has two manipulator arms and a motorized pan/tilt head carrying a color CCD camera as shown in Fig. 2.1. Each arm can move within 6 degrees of freedom. The whole system is controlled by a PC running XP which is responsible for controlling the two manipulator arms, any tools, the pan/tilt head, and also processing images from the CCD camera and other sensory information. The control program is written in C++.

In this paper only one of the robot arms was used. In the experiments we commanded the robot arm to move randomly at a fixed height above the table by driving joint 2 and joint 3 of the robot arm. After each movement, if the hand was in the current field of view of the camera, the eye system moved the camera to centre on the end of the robot finger, and then the pan/tilt head position (p, t) and current arm

Fig. 2.1 Experimental system for developmental coordination learning

joint values of the two joints used ($j2$, $j3$) were obtained to form a training set for the system; otherwise, if the hand tip is out of the view of the camera, this trial was ignored because the eye could not locate the arm end before setting up the mapping between pan/tilt and robot arm. After each trial, the obtained data ($p, t, j2, j3$) was used to train the mapping network, and this data was used only once. In order to simplify the image processing task of finding the end of the robot finger we marked the finger end with a blue cover. The position of the blue marker could be slid up and down the finger to effectively alter the length of the finger.

2.4.2 Constructive Learning and Adaptation in Tool-Use

To illustrate the network topological growth and parameter adjustments in constructive learning, Fig. 2.2 gives the structures of the hidden units at the $100th$ learning step and the $1597th$ learning step in eye/hand mapping. The results shows that at the beginning, the system used large neurons to quickly cover the whole space, and later on gradually built the details with smaller neurons when necessary, this let the system achieve more accuracy. This neuron growing process from coarse to fine using different neuron coverages is similar to infant development where the decrease in the size of neural receptive fields in the cortical areas relates to object recognition ability [24]. Figure 2.2 also demonstrates the changes of position and size of receptive field of each neuron. It should be noted that some neurons are removed in the learning process due to their small contribution to the sensory-motor mapping network.

Our next experiment was to test the network's adaptability to sudden changes in the motor-sensory relationship due to structural changes. We chose changes in

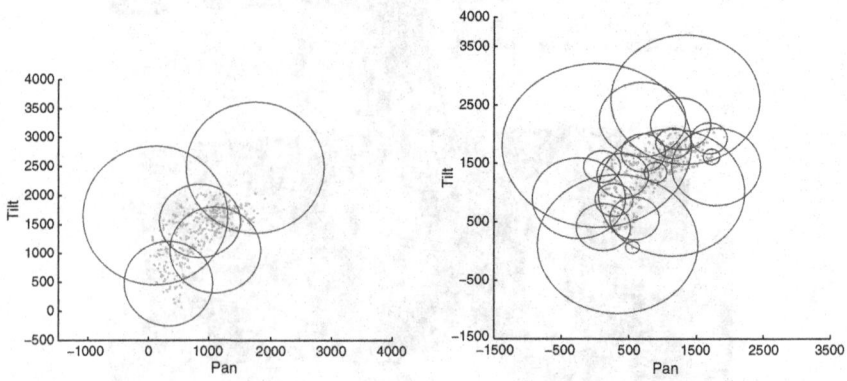

(a) Structure of hidden units at the $100th$ learning step

(b) Structure of hidden units at the $1597th$ learning step

Fig. 2.2 Distribution of the hidden units and their coverage in eye/hand mapping by RBF with SDEKF. The background points are the input learning points in the pan and tilt space of the camera head, and the circles are the hidden units of the eye/hand mapping network

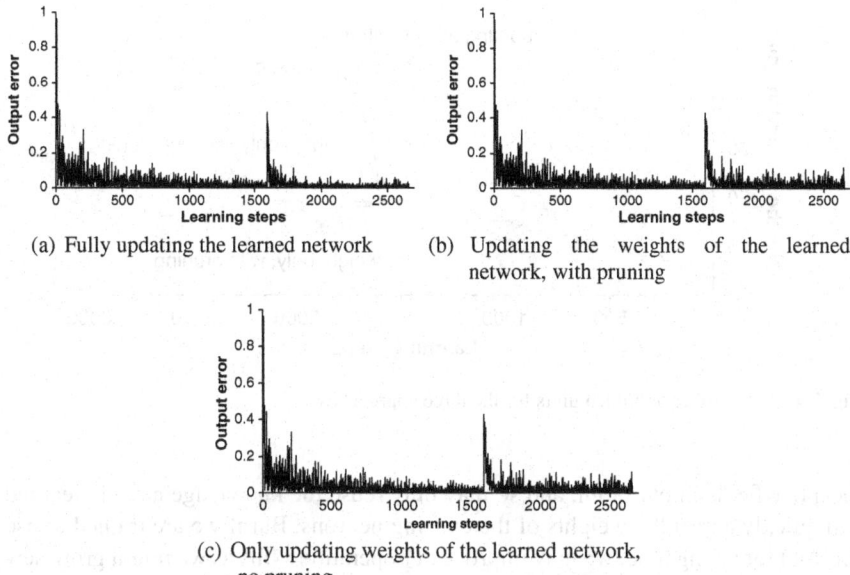

(a) Fully updating the learned network

(b) Updating the weights of the learned network, with pruning

(c) Only updating weights of the learned network, no pruning

Fig. 2.3 Adapting to structural change by reusing the learned network in different ways

finger length as a scenario to test this adaptability. Using a variety of tools with different sizes is necessary for a robot system to conduct different tasks, and the eye/hand mapping network's ability to quickly adapt to this change is crucial for the robot to re-gain its eye/hand coordination skills. We have tested three approaches to reusing and adjusting the learned eye/hand mapping network in order to re-gain coordination skills. As a test, at the 1598th trial (a purely arbitrary point) the finger length was changed in size from 27.5cm long to 20.5 cm long and we investigated the adaptation of the system to such a sudden change. Figure 2.3(a) shows the output error when all the parameters of the learned network are adjusted, including adding possible nodes, moving node centres, adjusting widths of each node, and updating the weights. Figure 2.3(b) and Fig. 2.3(c) show the results of only adjusting the weights and keeping the parameters of the hidden units unchanged, but Fig. 2.3(b) used a pruning procedure as described in Section 2.2.2 to remove the insignificant hidden units, while Fig. 2.3(c) kept the hidden unit structure completely unchanged. From the results, we can see that all three methods quickly adapt to the sudden change in finger size. The method of adjusting the full network parameters achieved the best result. Although the other two methods did not change the parameters of the hidden units of the learned network, they obtained reasonable small errors. It is important to note that, the third method, which completely reused the original hidden unit structure in the mapping network and only adjusted weights, achieved a quite similar result to the second method with pruning. This may be similar to the approach that adults adopt to handle tool changes. We can quickly adapt to structural changes with little effort, but during such short time-scales we cannot regenerate

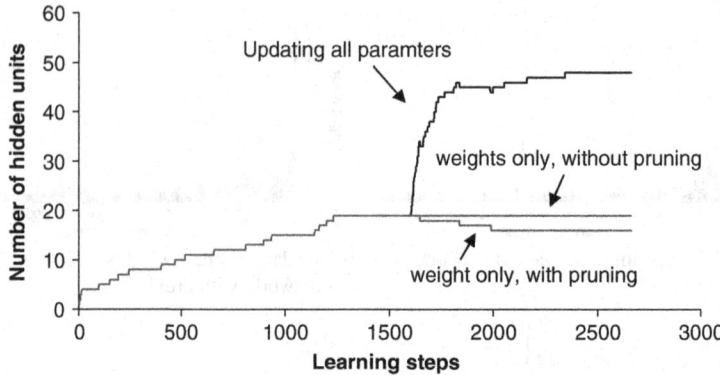

Fig. 2.4 The number of hidden units for the three approaches

receptive fields in our brain, and so may only reuse the knowledge already learned and quickly adjust the weights of the existing neurons. But if we are trained to use this tool for a long time, we may improve our operation skills as we might grow new neurons to support the changes as in Fig. 2.3(a).

Now considering network size, as shown in Fig. 2.4, the first method with full updating of all network parameters required by far the largest network, 48 nodes; while the second method removed three hidden units, reducing the network to 16 nodes; the third method kept the original network size, 19 nodes.

We have also studied the staged development in sensory-motor mapping learning process [7]. The system constructs sensory-motor schemas in terms of interlinked topological mappings of sensory-motor events, and demonstrates that the constructive learning moves to next stage if stable behaviour patterns emerges.

2.5   Conclusions

Constructive learning has advantages over static learning in sensory-motor mapping development for autonomous systems. It supports both topological network growth and parameter adjustments, which is supported by findings in psychology and neuroscience. It also has the advantage of adaptation to system changes such as in tool-use. A growing radial basis function network by constructive learning constructs the computational substrate for such sensory-motor mapping development. It forms a platform to examine the relationship between behaviour development and the growth of internal sensory-motor mapping network; the staged and developmental learning process through various constraints in motors and sensors; and active behaviour learning driven by intrinsic motivation. The experimental results on robot eye/hand coordination demonstrate the incremental growth of the mapping network and the system's adaptation to environmental changes.

References

1. Bousquet, O, Balakrishnan, K., and Honavar, V (1998). Is the hippocampus a Kalman filter? In *Pacific Symposium on Biocomputing*, pages 655–666, Hawaii.
2. Hihara, S., Notoya, T., Tanaka, M., Ichinose, S., Ojima, H., Obayashi, S., Fujii, N., and Iriki, A. (2006). Extension of corticocortical afferents into the anterior bank of the intraparietal sulcus by tool-use training in adult monkeys. *Neuropsychologia*, 44(13):2636–2646.
3. Hihara, S., Obayashi, S., Tanaka, M., and Iriki, A. (2003). Rapid learning of sequential tool use by macaque monkeys. *Physiology and Behavior*, 78:427–434.
4. Huys, Quentin JM, Zemel, Richard S, Natarajan, Rama, and Dayan, Peter (2007). Fast population coding. Neural Computation, 19(2):404–441.
5. Imamizu, H., Miyauchi, S., Tamada, T., Sasaki, Y., Takino, R., Puetz, B., Yoshioka, T., and Kawato, M. (2000). Human cerebellar activity reflecting an acquired internal model of a novel tool. *Nature*, 403:192–195.
6. Johnson-Frey, Scott H. (2004). The neural bases of complex tool use in humans. *Trends in Cognitive Science*, 8(2):71–78.
7. Lee, M.H., Meng, Q., and Chao, F. (2007). Developmental learning for autonomous robots. *Robotics and Autonomous Systems*, 55(9):750–759.
8. Lu, Yingwei, Sundararajan, N., and Saratchandran, P. (1998). Performance evaluation of a sequential minimal radial basis function (RBF) neural network learning algorithm. *IEEE Transactions on neural networks*, 9(2):308–318.
9. Maguire, Eleanor A., Gadian, David G., Johnsrude, Ingrid S., Goodd, Catriona D., Ashburner, John, Frackowiak, Richard S. J., and Frith, Christopher D. (2000). Navigation-related structural change in the hippocampi of taxi drivers. *PNAS*, 97(8):4398–4403.
10. Maravita, A. and Iriki, A. (2004). Tools for the body (schema). *Trends in Cognitive Science*, 8(2):79–86.
11. O'Keefe, J. (1989). Computations the hippocampus might perform. In Nadel, L., Cooper, L.A., Culicover, P., and Harnish, R.M., editors, *Neural connections, mental computation*. MIT Press, Cambridge, MA.
12. Piaget, Jean (1952). *The Origins of Intelligence in Children*. Norton, New York, NY.
13. Poggio, Tomaso (1990). A theory of how the brain might work. MIT AI. memo No. 1253.
14. Poggio, Tomaso and Girosi, Federico (1990). Networks for approximation and learning. *Proceedings of the IEEE*, 78(9):1481–1497.
15. Pouget, A. and Snyder, L.H. (2000). Computational approaches to sensorimotor transformations. *Nature Neuroscience supplement*, 3:1192–1198.
16. Quartz, S.R. and Sejnowski, T.J. (1997). The neural basis of cognitive development: A constructivist manifesto. *Brain and Behavioral Sciences*, 20:537–596.
17. Rao, R. and Ballard, D. (1997). Dynamic model of visual recognition predicts neural response properties in the visual cortex. *Neural Computation*, 9(4):721–763.
18. Rao, R. and Ballard, D. (1999). Predictive coding in the visual cortex. *Nature Neuroscience*, 2(1):79–87.
19. Shultz, T.R (2006). Constructive learning in the modeling of psychological development. In Munakata, Y. and Johnson, M.H., editors, *Processes of change in brain and cognitive development: Attention and performance XXI*, pages 61–86. Oxford: Oxford University Press.
20. Shultz, T.R., Mysore, S.P., and Quartz, S. R. (2007). Why let networks grow. In Mareschal, D., Sirois, S., Westermann, G., and Johnson, M.H., editors, *Neuroconstructivism: Perspectives and prospects*, volume 2, chapter 4, pages 65–98. Oxford: Oxford University Press.
21. Szirtes, Gábor, Póczos, Barnabás, and Lőrincz, András (2005). Neural Kalman filter. *Neurocomputing*, 65–66:349–355.
22. Todorov, E. and Jordan, M.I. (2002). Optimal feedback control as a theory of motor coordination. *Nature Neuroscience*, 5(11):1226–1235.

23. Weng, Juyang, McClelland, James, Pentland, Alex, Sporns, Olaf, Stockman, Ida, Sur, Mriganka, and Thelen, Esther (2001). Autonomous mental development by robots and animals. *Science*, 291(5504):599–600.
24. Westermann, G. and Mareschal, D. (2004). From parts to wholes: Mechanisms of development in infant visual object processing. *Infancy*, 5(2):131–151.

Chapter 3
Using Digital Watermarking for Securing Next Generation Media Broadcasts

Dominik Birk and Seán Gaines

Abstract The Internet presents a problem for the protection of intellectual property. Those who create content must be adequately compensated for the use of their works. Rights agencies who monitor the use of these works exist in many jurisdictions. In the traditional broadcast environment this monitoring is a difficult task. With Internet Protocol Television (IPTV) and Next Generation Networks (NGN) this situation is further complicated.

In this work we focus on Digitally Watermarking next generation media broadcasts. We present a framework which provides the ability to monitor media broadcasts that also utilises a Public Key Infrastructure (PKI) and Digital Certificates. Furthermore, the concept of an independent monitoring agency, that would operate the framework and act as an arbiter, is introduced. We finally evaluate appropriate short signature schemes, suitable Watermarking algorithms and Watermark robustness.

Keywords Next generation networks · Broadcast monitoring · Public key watermarking · IPTV · PKI · Short signature

3.1 Introduction

Radio and television are audio and video broadcasting services typically broadcasted over the air, cable networks or satellite networks. Since the advent of the Internet, the distribution of media content has always been a principal goal, however for many years this was not realised due to the prohibitive cost and limited capabilities of personal computers.

D. Birk (✉)
Horst Görtz Institute for IT Security, Ruhr-University Bochum, Building IC 4, D-44780, Bochum, Germany
e-mail: dominik@code-foundation.de

S. Gaines
VICOMTech Research Center, Paseo Mikeletegi 57, E-20006, San Sebastian, Spain

S.-I. Ao et al. (eds.), *Advances in Machine Learning and Data Analysis*,
Lecture Notes in Electrical Engineering 48, DOI 10.1007/978-90-481-3177-8_3,
© Springer Science+Business Media B.V. 2010

With the transition to digital media streams received over the Internet, new challenges loom. Today, the practices of recording, distribution and copying multimedia content is easy and straightforward [11]. Due to these facts, it is more and more difficult to enforce copyright and to safeguard intellectual property for broadcast media.

Digital Watermarking [6], which may be considered as a form of steganography [8], attempts to address this problem by embedding information within the digital signal. The embedded Watermark is invisible to the user, should not affect the perceived aesthetic quality of the final signal, nor should the Watermark reveal any clues about the technique used to embed it. However, it is debatable whether traditional Watermarking systems, which are based on disclosure of the key needed to embed and to detect the watermark are generally suitable for proving ownership or authentication. Therefore, we established a framework based on asymmetric public-key cryptography which is used for exhaustive authentication with the help of *Blind Watermarking* techniques.

In many jurisdictions broadcasters have regulatory obligations which attempt to protect the intellectual property [8] and copyrights of authors, songwriters, performers, actors, publishers, etc. Furthermore, in some jurisdictions there exists bodies charged with the defense of the rights of intellectual property and copyright holders and the calculation, charging and collection of performance royalties on the use of these protected works. Currently, there are several cases in which broadcasters cannot confidentially confirm that their royalties liabilities are correctly calculated. This is because they currently do not employ a viable automated system to measure what protected works are broadcasted, how often and when. Therefore a gap has opened up in the actual amount charged by the rights bodies and the correct payable royalties liability of the broadcaster.

This paper describes a specific Watermarking concept that may be used for identifying next generation media broadcast streams based on PKI authentication. We introduce a formal PKI framework in section 3.3 allocating authentication methods and then focus on procedures and measures for Watermarking media streams in legacy networks as well as NGNs using PKI. We prove our proposal through an exemplified scenario on video stream Watermarking.

3.2   Framework Overview

A general overview of the framework with its three parties can be seen in Fig. 3.1. It makes use of two additional frameworks, the PKI and the Watermarking Framework.

The PKI Framework, described in chapter 3.3, is used for establishing a trust network between all of the involved entities. The *broadcaster* (BC) can initialise the monitoring process for metering his use of protected works and hence the royalties payable rights entity can also launch the monitoring process for billing purposes. In practice, the PKI procedures (3.1) should be established as the first step in the

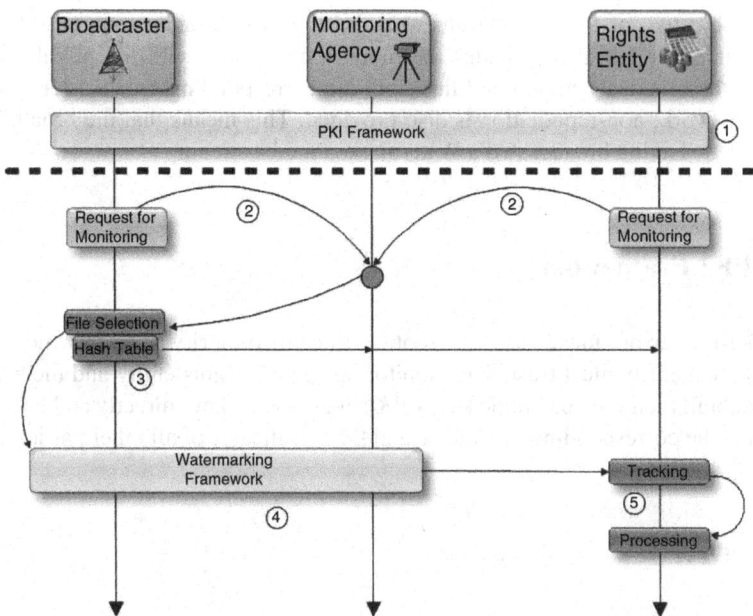

Fig. 3.1 General Framework Overview

deployment of the framework. The PKI is necessary for establishing a trusted relationship with the purpose of distributing authenticated private and public keys utilising digital certificates. To start the process of monitoring, a "Request for Monitoring" (3.2) is sent to the *monitoring agency* (MA).

Afterwards, the broadcaster selects a piece of content which he wants to stream (3.3) and computes the corresponding hash table (see Section 3.4.2.3). This hash table is carried over a secure and authenticated channel to the MA as well as to the *rights entity* (EX). Afterwards, the broadcaster initiates the process defined by the Watermarking Framework.

The Watermarking Framework specifies procedures for Watermark embedding, retrieval and verification in media streams (3.4). The rights entity is the entity which charges broadcasters for distributing media content over a chosen distribution network. It also attempts to track and process broadcast media with the help of information obtained by the monitoring agency. The broadcaster will sign the stream which is about to be broadcasted with his private key. Subsequently, the corresponding signature will be embedded into the media stream with known Watermarking techniques. Later in the process, the monitoring agency will extract the Watermark and verify the signature. So, the agency can be sure that only the original broadcaster broadcasted the media content, due to the fact that additional security metadata, such as timestamps and identifiers, are used. Additionally, EX can also verify the signature in order to prevent abuse by the MA (5).

The objective of the whole framework is to let the broadcaster mark the file stream uniquely but also provides the monitoring agency with the possibility to identify the broadcast stream and therefore the corresponding broadcaster. Within this framework, non-repudiation is also provided. This means that the broadcaster cannot deny having broadcasted a Watermarked media stream.

3.3   PKI Framework

The PKI framework makes use of a root Certificate Authority (CA) in which each participating entity must trust. The monitoring agency, rights entity and the broadcaster submit their created public keys (PK) or create the keys directly at the CA for receiving the corresponding Certificate and the Certificates of all other participants.

3.3.1   Overview

Trust forms the basis of all communication, be it physical or electronic. In the case of electronic communication, building trust is quite difficult as the identity of the other entity remains concealed. While a PKI normally provides confidentiality, non-repudiation, authentication and integrity, our framework mainly focuses on authentication and non-repudiation.

A detailed description of the three stages in Fig. 3.2 will be given in the following section.

1. This step demonstrates a particular need of a PKI. The public key (PK_{CA}) of the CA has to be pre-distributed in an authenticated manner to any involved entity, otherwise no secure communication with the CA is possible.
2. As soon as the entities have received the authenticated PK_{CA} over a secure, authenticated channel, they create their own secret (SK_X, SK_{MA} and SK_{BC}) and public key (PK_X, PK_{MA} and PK_{BC}) and subsequently send a PKCS#10 Certificate request to the Certificate Authority. With Digital Signatures in Certificate requests, the CA can be sure that the sender has a private key related to the public key. Therefore, the sender has a proof of possession [4] but the receiver needs to assure that the entity with which he is communicating is not spoofed.
3. If the CA receives the Certification request, it behaves like a Registration Authority (RA) and tries to validate the information stored in the PKCS#10 file. If it is valid, a X.509 Certificate is issued by signing the corresponding PK of the entity. Afterwards, all Certificates are distributed to all entities for authentication reasons. So, the broadcaster owns now a Certificate concerning the PK of EX which was issued by the corresponding CA. The Certificate will be used during the Watermarking processes in order to authenticate the sender.

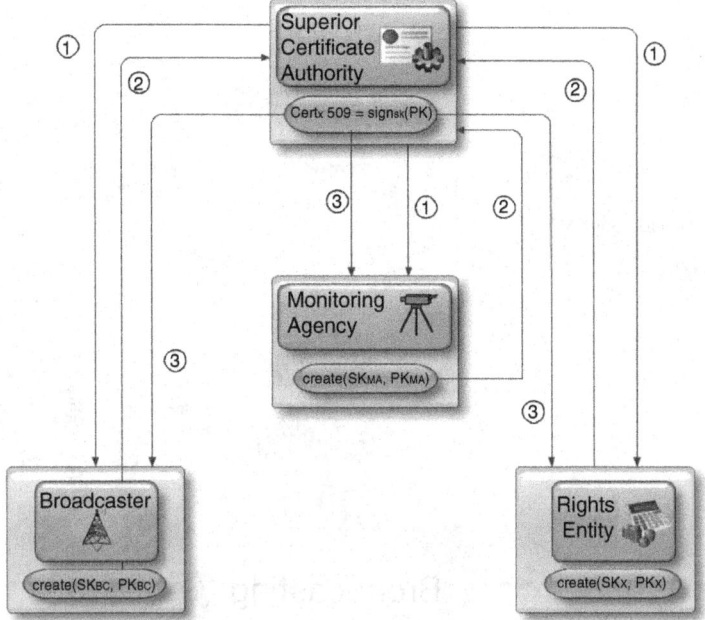

Fig. 3.2 PKI Framework Overview

As previously mentioned, the main purpose of these protocol steps is providing full authentication. Now, the broadcasting entity could sign a message and send it to the monitoring agency or indeed to the rights entity and both entities could be assured, that the message was sent by the broadcaster.

3.4 Watermarking Framework

The Watermarking Framework, illustrated in Fig. 3.3, specifies the communication protocol between the broadcaster and the monitoring agency in which the rights entity is not involved. Furthermore, the Watermarking Framework provides a detailed insight into procedures for creating, detecting, extracting and verifying the Watermark.

3.4.1 Overview

The framework is initialised at the moment the broadcaster had chosen a content file and transferred the corresponding hash table to the MA (see Algorithm 3.4.2.3). Afterwards, no further information needs to be sent to the MA due to the use of

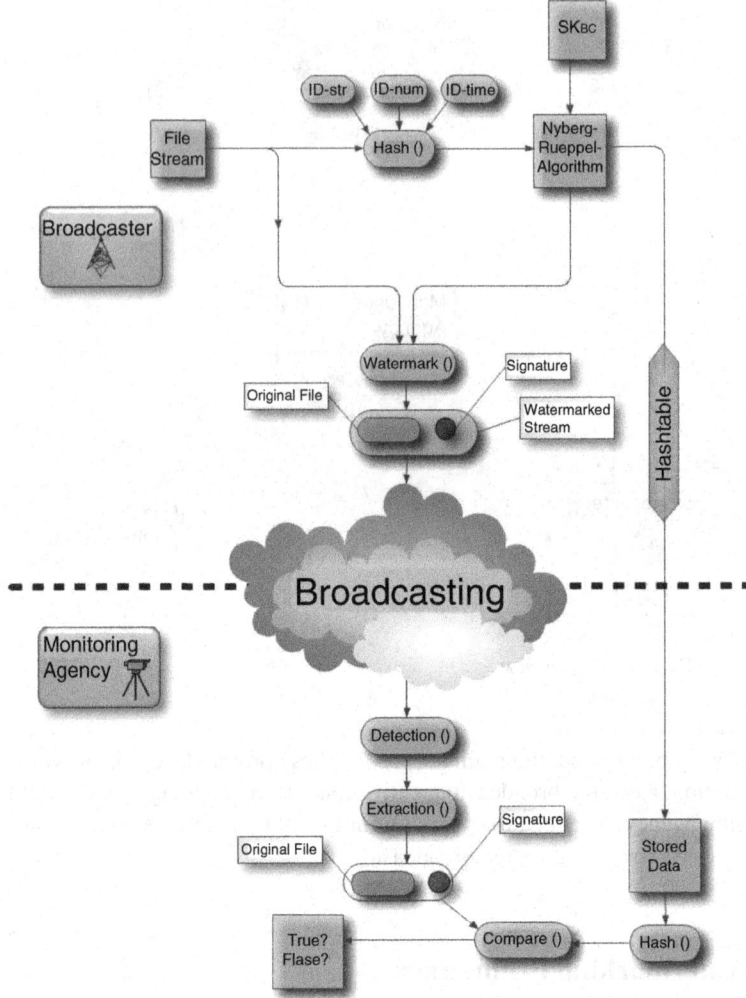

Fig. 3.3 Watermarking Framework Overview

message recovering signatures. So the MA can be sure about who broadcast the stream and what stream has been broadcasted. This information is sufficient for metering and charging purposes.

The chief characteristic of a Watermarking scheme for copyright protection, or DRM, is that the Watermark cannot be separated from the medium without knowledge of a secret value. We, in our specific case, target on another characteristic: sender authentication. It should be possible to identify the broadcasting station unambiguously and show exactly who broadcasted what stream and when.

Therefore, our Watermark information contains a digital signature issued by the broadcaster. Each entity that receives the broadcast stream and owns the corresponding broadcaster certificate, can clearly verify the distributed stream with the help of the corresponding PK.

Below, we discuss suitable signature schemes and Watermarking algorithms. We introduce adequate procedures for embedding and retrieving the Watermark with the help of a beacon in addition to verifying the signature.

3.4.2   Signature Schemes

A principal requirement to all Watermarking systems is the need for a small Watermark. The larger the Watermark, the larger are the chances for adversely affecting the quality of the streamed media. In our case, the Watermark depends on the corresponding signature which has to authenticate the media stream. Interference and transaction defects could cause problems in extracting the Watermark. Therefore, the signature scheme output has to be as small [9] as is possible to be able to embed the Watermark as often as possible and to be repeated multiple times throughout the stream.

The Nyberg-Rueppel ([10], hereafter NR) signature scheme focuses on the size of the input and output and is a DSA-like signature with message recovery.

3.4.2.1   The Nyberg-Rueppel Signature Scheme

NR is perfectly suited to messages shorter than ten bytes but leaves the question of dealing with short messages, of say fifteen bytes, unanswered. In our specific case, the hash to be signed is exactly 10 bytes long and brings only a marginal risk of collision (see Section 3.4.2.2 for further details).

Message recovery [1], another characteristic of NR signatures, provides means so that the original message can be extracted out of the signature. In our given case, this characteristic aligns with our goals. The hash value of the message m does not need to be created by the monitoring agency, due to the fact that it can be extracted from the signature due to the aforementioned message recovery characteristic. However, it is necessary to transfer a hash table (see Section 3.4.2.3) once from the BC to the MA. This could happen in periodical time-frames.

The complete NR algorithm is shown in Algorithm 3.1 using the standard discrete logarithm (DL) problem.

3.4.2.2   Short Hash Methods

Hash functions are often used in digital signature algorithms. The message m that is about to be hashed, in our case, consists of an identifier string ID-str concatenated with an ID number ID-num and an unique times-tamp ID-time:

Algorithm 3.1 Nyberg-Rueppel signature generation and verification

---

Summary: the broadcaster signs a message $m \in \mathcal{M}$. The monitoring agency can verify the broadcaster's signature and recover the message m from the signature.

1. *Signature Generation*. The broadcaster has to do the following:

   (a) Compute $\tilde{m} = R(m)$.
   (b) Select a random secret integer k, $1 \leq k \leq q - 1$ and compute $r = \alpha^{-k} \bmod p$.
   (c) Compute $e = \tilde{m} r \bmod p$
   (d) Compute $s = ae + k \bmod q$.
   (e) The broadcaster's signature for the specific m is the pair (e, s).

2. *Verification*. To verify the broadcaster's signature (e, s) on m, the monitoring agency should do the following:

   (a) Obtain the broadcaster's authentic public key (p, q, α, y) and verify it with the corresponding certificate delivered by the CA earlier (see Fig. 3.2).
   (b) Verify that $0 < e < p$; if not, reject the signature.
   (c) Verify that $0 \leq s < q$; if not, reject the signature.
   (d) Compute $v = \alpha^s y^{-e} \bmod p$ and $\tilde{m} = ve \bmod p$
   (e) Verify that $\tilde{m} \in \mathcal{M_R}$; if $\tilde{m} \notin \mathcal{M_R}$ then reject the signature.
   (f) Recover $m = R^{-1}(\tilde{m})$.

$$m = \textit{ID-str} \parallel \textit{ID-num} \parallel \textit{ID-time} \tag{3.1}$$

The ID-str could be represented through the name of the media content, for instance. The ID-num could be an identification number. The ID-time is a unique time-stamp which prevents replay-attacks. This means, that an adversary may not record the stream and broadcast it later again on an authorised channel which is also monitored.

3.4.2.3 Hash Table for Verification Purposes

A hash table ht in our specific case is a data structure that associates the hash value with ID-str, ID-num and ID-time. The hash table contains several important attributes and is essential for the verification process by the MA (see Fig. 3.4).

Transferring the hash table to the MA, can be compared to the cryptographical *commitment scheme*, visualized in Algorithm 3.2, except that the hash table has no hidden value. The prover, respectively the BC, sends the "commitment" in form of the hash table ht to the verifier (the MA). MA will forward the signed hash

Hash	ID-str	ID-num	ID-time
e80c78299acc041ffd23	Title A	42133	34523312
9a2002a978b5c7538952	Title B	87565	56245323
65ae7da24e501c95a0ae	Title C	52332	6345231

Fig. 3.4 Structure of the Hash table

Algorithm 3.2 Secure and authentic Hash table distribution

Summary: during the *commitment phase*, the hash table is transferred to MA and EX. During the *opening phase*, BC proves to MA and EX that he is broadcasting one of the items in the hash table.

1. *commitment phase*:

   ---
   1. BC $\longrightarrow$ MA: $\text{enc}_{PK_{MA}}(\text{sign}_{SK_{BC}}(ht))$

   2. MA $\longrightarrow$ EX: $\text{enc}_{PK_{EX}}(\text{sign}_{SK_{BC}}(ht))$
   ---

2. *opening phase*:

   ---
   1. BC $\longrightarrow$ MA: watermark($\text{sign}_{SK_{BC}}(hs)$)

   2. MA: extract *signature* from stream with the help of beacon

   3. MA $\longrightarrow$ EX: $\text{enc}_{PK_{EX}}(\text{sign}_{SK_{BC}}(hs))$

---

table to the rights entity but encrypts it with the corresponding PK_{EX} in order to guarantee secrecy which is needed to prevent other parties from viewing the hash table. This can be seen as the *commitment phase* and takes place directly after having chosen the file to be streamed. The encryption is necessary due to the possibility that the hash table of a potential business rival might be seen by another party. Later, after broadcasting the media content, the verifier can scrutinise, with the help of the message recovery characteristic of the signature, whether the BC broadcast the content correctly or not (*opening phase*). As can be seen in Fig. 3.4 that the hash value is only made of 10 bytes/20 hex characters. For verification, the MA needs to extract the hash value out of the signature with the help of the message recovery characteristic. Afterwards, he will look up the hash in the transferred hash table and check whether the corresponding ID-fields are valid. The same procedures can be done by the EX in order to be sure that the MA is not cheating.

For instance, if a video stream was recorded and replayed at a later point in time, the MA will recognise that due to this fact it will not match the ID-time field in the hash table. A video stream can only be validly broadcast once. If the BC tries to cheat by changing the ID-str field for a piece of media content with a lower or no payable royalty, the MA will detect that.

3.4.2.4   Case Study: Video Broadcaster

The *Internet Movie Database (IMDB)* published interfaces for several systems to access the IMDB locally. For our case study, we downloaded the complete IMDB title textfile which contains currently 1.206.730 different movie titles. We used the movie title as a ID-str and created a unique number used as the ID-num. The time-stamp ID-time was the current date parceled as a unixtimestamp. An example of an assignment between unixtimestamp and normal time can be seen in Eq. 3.2.

$$05/07/1982@00:00 \implies 389592000 \tag{3.2}$$

For instance, in our simulation, m looked like this:

$$m = \underbrace{\text{Title A}}_{\textbf{ID-str}} \;||\; \underbrace{23754}_{\textbf{ID-num}} \;||\; \underbrace{534056}_{\textbf{ID-time}} \tag{3.3}$$

We created 1,206,730 different messages m and subsequently hashed them with MD5 and SHA-1. Afterwards, we extracted the first 10 bytes which satisfy the first 20 characters of the output HEX value. No collisions were detected for both hash functions, MD5 and SHA-1, even with only using the first 10 bytes of the hash-sum.

$$hs = [0...9]hash(m) \tag{3.4}$$

3.4.3 Suitable Watermarking Algorithms

A substantial body of research in Watermarking algorithms can be found in literature [12]. However, in our specific case, the Watermark should have special control characteristics which are required to guarantee the ability to verify the embedded signature by the monitoring agency.

- **Robustness**
  Robust Watermarks are designed to resist against heterogeneous manipulations and therefore not substantial for our framework [3]. Our framework focuses on authentication, not on robustness against manipulation. Only robustness against accidental manipulation or signal interference would be useful in our case.
- **Invisibility**
  The Watermark embedded into a video stream should be visually imperceptible.
- **Inaudibility**
  The Watermark of an audio stream or the audio track in a video stream should be unaudible.
- **Complexity**
  Watermarking and Watermark retrieval should, in principle, have low complexity. Due to the fact that our case focuses on streaming applications, the functions for embedding and retrieving of the Watermark should be as simple as possible so that on the fly, or faster than realtime, Watermarking is possible.
- **Compressed domain processing**
  We assume that the broadcaster will store the media files in a compressed format. Referring to the above complexity requirement, embedding the watermark into the compressed video stream should be possible, specific decode and recode steps for watermarking are undesirable as not to affect the overall performance of the system.

In our case, there is no need to keep the Watermark private. Each participant in the framework may extract the Watermark via the known *beacon* needed to locate the Watermark. Afterwards, the signature can be extracted from the Watermark and verified with the help of the corresponding public key.

3.4.3.1   Proposed Watermarking Algorithm

Basically, a Watermarking system for our purposes can be described by a tuple $\langle \mathcal{O}, \mathcal{S}, \mathcal{W}, \mathcal{H}, \mathcal{P}, \mathcal{G}, C_S, E_H, D_H, V_P \rangle$ where $\mathcal{O}$ is the set of all original data, a video stream for instance. The set $\mathcal{S}$ contains all secret keys needed for creating an unforgeable signature. $\mathcal{W}$ represents the set of all Watermarks (signatures, in our case) and $\mathcal{H}$ the set of all beacons. Beacons in our scenario are markers that signify the presence and start of a Watermark bit sequence in the signal. The beacon substitutes the key in normal Watermarking systems. $\mathcal{P}$ describes the set of public keys which are needed to verify the signature and $\mathcal{G}$ represents the set of Certificates issued by the CA.

Four functions are described as followed:

$$C_S : \mathcal{O} \times \mathcal{S} \longrightarrow \mathcal{O} \tag{3.5}$$

$$E_H : \mathcal{O} \times \mathcal{S} \times \mathcal{W} \times \mathcal{H} \longrightarrow \mathcal{O} \tag{3.6}$$

$$D_H : \mathcal{O} \times \mathcal{H} \longrightarrow \mathcal{W} \tag{3.7}$$

$$V_P : \mathcal{W} \times \mathcal{P} \times \mathcal{G} \longrightarrow \{1, 0\} \tag{3.8}$$

C_S focuses on creating the corresponding Watermark through a signature. E_H describes the function for embedding the Watermark and D_H respectively the function for extracting it. Furthermore, V_P stands for the verification function needed to check if the Watermark is valid.

The Watermark w is created with

$$w = sign_{SK_{BC}}(hs) \tag{3.9}$$

and outputs a short bit-string which contains the signature of the reduced hash-sum. See Eq. 3.4 for further details about the reduced hash-sum hs.

3.4.4   Embedding the Watermark

In this subsection we focus on the embedding process of the signature/ Watermark. Hartung and Girod proposed in 1998 [7] a method which focuses on Watermarking MPEG-2 video data. We adopt the proposed methods for our purposes of embedding the signature into a given video broadcast stream. For further information, the interested reader is referred to [7].

The basic concept of Hartung and Girod [7] was to present a Watermarking scheme for MPEG-2 encoded as well as uncompressed video based on spread-spectrum methods [5].

Let

$$a_j, \quad a_j \in \{-1, 1\}, \quad j \in \mathbf{N} \tag{3.10}$$

be the Watermark bit sequence to be hidden in a linearised video stream. In our case, this bit sequence contains the signature which was created by signing the reduced hash with a specific short signature method based on the NR algorithm (see Algorithm 3.1). This discrete signal is up-sampled by a factor cr called the chip-rate, to obtain a sequence

$$b_i = a_j, \quad j \cdot cr \le i < (j + 1) \cdot cr, \quad i \in \mathbf{N} \tag{3.11}$$

so as to provide redundancy. The new bit sequence b_i is modulated by a pseudo-noise signal, respectively the beacon in our specific case, p_i whereas $p_i \in \{-1, 1\}, i \in \mathbf{N}$ and previously scaled by a constant $\alpha_i \ge 0$. Therefore, the spread spectrum Watermark now consists of

$$w_i = \alpha_i \cdot b_i \cdot p_i \quad i \in \mathbf{N} \tag{3.12}$$

Afterwards, the spread spectrum Watermark w_i is added to the line-scanned digital video signal v_i yielding the new, Watermarked video signal

$$\tilde{v} = v_i + w_i = v_i + \alpha_i \cdot b_i \cdot p_i \tag{3.13}$$

Due to the noisy appearance of p_i, the spread spectrum watermark w_i is also noise-like and therefore difficult to detect and remove.

In ordinary Watermarking schemes, p_i is typified as the secret key. As already noticed, our proposed scheme doesn't need a secret key, therefore the p_i sequence is a beacon known to both the broadcaster and the monitoring agency. Hartung and Girod proposed to create p_i with the help of feed-back shift registers producing m-sequences or chaotic physical processes. However we propose to use a public and arranged sequence which does not itself cause interference itself and is frequently repeated.

3.4.5 Retrieval of the Watermark

The proposed methods rely on *Blind Watermarking* techniques and therefore do not need the original video stream in the retrieval process. To correctly decode the information, the "secret" beacon p_i must be known. Due to the public nature of p_i in our case, the monitoring agency knows the sequence.

Optionally, the Watermarked stream can be high-pass filtered in order to improve the performance of the overall Watermarking system. Afterwards, the possibly Watermarked video stream $\bar{v}$ is multiplied by the same noise-like beacon stream p_i that was used in the embedding process.

$$
s_j = \underbrace{\sum_{j \cdot cr \leq i < (j+1) \cdot cr} p_i \cdot \bar{\bar{v}} = \sum_{j \cdot cr \leq i < (j+1) \cdot cr} p_i \cdot \bar{v}}_{\Sigma_1}
$$

$$
+ \underbrace{\sum \sum_{j \cdot cr \leq i < (j+1) \cdot cr} p_i \cdot \overline{p_i \cdot \alpha_i \cdot b_i}}_{\Sigma_2} \approx \sum_{j \cdot cr \leq i < (j+1) \cdot cr} p_i^2 \cdot \alpha \cdot b_i \quad (3.14)
$$

We now assume, in accordance with Hartung and Girod, that Σ_1 is zero due to the fact that the video signal has been filtered out. Furthermore, we assume that $p_i \cdot \alpha_i \cdot b_i \approx p_i \cdot \alpha_i \cdot b_i$ which therefore means, that the high-pass filtering has negligible influence on the white pseudo-noise Watermark signal. Following the proposal of Hartung and Girod, the sign of the correlation sum is the embedded information bit

$$
sign(s_j) = sign(a_j) = a_j \quad (3.15)
$$

3.4.6   Verifying the Signature

It is possible for the monitoring agency to verify the signature which is represented by the extracted bit sequence. The method V_P verifies the signature with the help of the corresponding public key which is taken from the Certificate in order to be sure, that only the public key belonging to the correct broadcaster is used.

But verification alone is not sufficient in our case. The monitoring agency has to be sure, that the broadcaster keeps to the preassigned streaming plan. Therefore, a signature algorithm was used which has the characteristic called message recovery. This means, the message can be extracted from the signature after verifying it. See the Algorithm 3.1 for more details about the verification procedure.

Due to the fact, that the broadcaster signed a hash of m (see Section 3.3), the monitoring agency has to look up the hash in his hash table transferred to him beforehand (Algorithm 3.2). So he can be sure that the extracted and verified hash belongs to the correct broadcaster. Furthermore, the MA can be sure, that the broadcaster broadcasted the video stream in a given time-frame by comparing the embedded time-frame ID-time in m. If the time-stamp is not within a 60 time-frame, it could be possible, that someone attempted to replay the recorded video stream. It might also be possible, that is is not the original broadcaster is trying to cheat, but that another broadcaster has recorded the stream and is attempting to stream/broadcast it.

3.5 Conclusions and Future Work

The schemes proposed in this paper may be viewed as attractive to both broadcasters and rights agencies. This model provides the broadcaster and the rights entity with an automated and trust worthy method for measuring the exploitation of protected works. The paper introduces the concept of an independent third party that monitors and balances the interests of the broadcaster and rights entity.

We discuss the new technologies and distribution models faced by the entertainment and broadcasting sectors. We evaluate established short signature schemes, such as Nyberg-Rueppel, that could be integrated into a final system.

Boneh et al. [2] proposed a public key encryption scheme in which there is one public encryption key, but many private decryption keys. This scheme could be used if multimedia content should be encrypted and distributed over given channels. However, in our specific case, we do not focus on encrypting the content. The goal is authentication and non-repudiation for the broadcaster, so that the MA is able to uniquely identify the sender.

Though, a similar schmeme could be used for the same purposes. Due to the fact, that only one unique public key exists, but many corresponding private keys, the broadcaster could encrypt a secret value with this public key and put the ciphertext into the media stream. This has several advantages. The MA as well as the EX could obtain a private key for the ciphertext ans used to decrypt the content. This means that more than one corresponding private key could be used for different monitoring agencies. In addition, authentication is also given in this context. Due to the fact that there is only one public key which is obtained by the broadcaster, only this broadcaster could encrypt the secret value. But this postulates that the public key is not "public" in general.

References

1. Ateniese, G. and de Medeiros, B. (1999). A signature scheme with message recovery as secure as discrete logarithm.
2. Boneh, Dan and Franklin, Matthew (1999). An efficient public key traitor tracing scheme. pages 338–353. Springer-Verlag.
3. Chen, B and Wornell, G W (1999). Provably robust digital watermarking. In *in Proc. SPIE Multimedia Systems and Applications II*, pages 43–54.
4. Choudhury, Suranjan (2002). *Public Key Infrastructure and Implementation and Design*, volume 1. Wiley & Sons.
5. Cox, Ingemar, Kilian, Joe, Leighton, Tom, and Shamoon, Talal (1997). Secure spread spectrum watermarking for multimedia. *IEEE Transactions on Image Processing*, 6(12):1673–1687.
6. Coxy, Ingemar J., Kiliany, Joe, Leightonz, Tom, and Shamoony, Talal (1996). A secure, robust watermark for multimedia.
7. Hartung, Frank and Girod, Bernd (1998). Watermarking of uncompressed and compressed video. *Signal Processing*, 66(3):283–301.
8. Lu, Chun-Shien (2005). *Multimedia Security: Steganography and Digital Watermarking Techniques for Protection of Intellectual Property*. IDEA GROUP PUBLISHING.

 9. Naccache, David and Stern, Jacques (2001). Signing on a postcard. *Lecture Notes in Computer Science*, 1962:121–123.
10. Nyberg, Kaisa and Rueppel, Rainer A. (1993). A new signature scheme based on the dsa giving message recovery. In *Proceedings of the 1st ACM CCCS*, Fairfax. ACM.
11. Peitz, M and Waelbroeck, P (2003). Piracy of digital products: A critical review of the economics literature. cesifo working paper series no. *Information Economics and Policy*, (1071):2003.
12. Seitz, Juergen (2005). *Digital Watermarking for Digital Media*. Information Science Publishing.

Chapter 4
A Reduced-Dimension Processor Model
Incorporating Microarchitectural Parameters and Software's Dynamic Characteristics

Azam Beg

Abstract Architectural simulators used for microprocessor design study and optimization can require large amount of computational time and/or resources. In such cases, models can be a fast alternative to lengthy simulations, and can help reach a designer near-optimal system configuration. However, the non-linear characteristics of a processor system make the modeling task quite challenging. The models not only need to incorporate the micro-architectural parameters but also the dynamic behavior of programs. This paper presents a hybrid (hardware/software), non-linear model for processors. The model provides accurate predictions of processor throughput for a wide range of design space. We used different groups of code basic blocks to investigate their relationships to the execution efficiency of a superscalar processor. For this purpose, we utilized the frequencies of the blocks to represent runtime nature of ten benchmark programs. We were able to reduce the number of hardware and software parameters by employing correlation coefficients and principal component analysis.

Keywords Processor throughput prediction · Processor model · Instructions per cycle (IPC) · Micro-architecture simulation · Code basic blocks · Software dynamic behavior

4.1 Introduction

Microarchitectural features of processor tend to become more and more complex in the newer designs. Using simulators to investigate the design space also takes longer when design complexity increases, because the detailed, cycle-accurate simulators take longer to run. The time needed to explore different design options can be reduced by using performance prediction models [1]. The models can be

A. Beg (✉)
College of Information Technology, United Arab Emirates University, Al-Ain, UAE
e-mail: abeg@uaeu.ac.ae

S.-I. Ao et al. (eds.), *Advances in Machine Learning and Data Analysis*,
Lecture Notes in Electrical Engineering 48, DOI 10.1007/978-90-481-3177-8_4,
© Springer Science+Business Media B.V. 2010

mathematical, statistical, or machine learnt in nature. Most of these models utilize only the hardware parameters [2–4]; the models do not include any parameters dependent on the programs. It is quite obvious that a processor's execution efficiency is affected not just by the static characteristics of a program but also by the dynamic (runtime) characteristics (Static features of programs include number of code lines, etc.) The runtime performance can be affected greatly by the programs' basic block sizes and their execution frequencies. Basic blocks are sets of instructions that contain a single control instruction, such as a conditional or a non-conditional branch. The control instruction is the last instruction of a basic block. (The terms 'basic code block,' 'basic block,' and 'block' are used alternatively in this paper.)

4.1.1 Paper Outline

In this paper, we present *neural network* (NN) based prediction models for a superscalar processor's performance; the performance is measured in terms of *instructions completed per cycle* (IPC). The model's inputs include both hardware and (runtime dependent) software parameters. The 'hardware' inputs to the model include microarchitectural features such as fetch, decode, issue, and commit widths; number of integer and floating point ALUs; etc. The 'software' inputs are mainly the sets of basic block frequencies (instead of single input value, i.e., the average block size [5, 6]). Inclusion of multiple software parameters helps one represent different programs distinctly. Unlike our previous work [1], which used an exhaustive set of hardware and software parameters, in this paper, we use a *reduced* set. Two separate techniques, namely, correlation coefficients and principal component analysis (PCA), were used to evaluate different reduced sets.

The paper is organized as follows: We first present related work and background information. Then we provide the particulars of the experimental setup and the data processing details. These are followed by the NN implementation and usage discussion. Lastly, we provide our conclusions and the direction of our future work.

4.1.2 Related Research

Hardware development is traditionally expedited using software models. The models are implemented in high level or hardware description languages. Relevant benchmark programs are then run on the models to get good approximations of actual hardware. A processor system model needs to be inclusive of both the program and hardware behavior. The program behavior can be *dynamic* or *static*. The dynamic characterization can be done by capturing repeating patterns in a program [7, 8] Cycle-accurate simulators tend to be accurate but may require weeks of simulation time with programs running for a few billion cycles [3]. Noonburg and Shen [4] utilized the benchmark (program) traces to create a model for superscalar instruction level parallelism (ILP). They combined the ILP and hardware parameters in their model. The prediction error with their models was as high as 22% for some of the SPEC CPU95 benchmarks [9]. Wallace and Bagherzadeh [2], and Hossain et al. [3] presented models for conventional and trace caches, respectively. The analytical model by Hossain et al., due to its limited scope (only the caches, and not the full processor) had higher prediction accuracy – 7% to be specific. Different parts of a program can be steady state or cyclical in nature; this property of programs was exploited in Hamerly et al.'s simulation tool [7]. To speed up simulation, Wunderlich et al. [10] statistically characterized the full-length benchmarks into smaller subsets. Joseph et al. [11] collected performance measures from detailed simulations and then used radial basis functions (RBFs) to build a model as an alternative to simulations. Their model provided cycles per instruction (CPI) estimates with error ranges of 1.5–12% for one of the SPEC CPU2000 benchmarks [12], and 1.5–23% for another. Lee et al.'s simulator [14] identified basic blocks that repeated often; it then used block behavior within a combination of *SimpleScalar*'s [13] fast (*sim-cache*) and slow (*sim-outorder*) modules to speed up the simulations by a factor of 3.3. A similar method of capturing the dynamic nature of a program is to detect its recurring patterns (called *program phases*). This approach requires one to pick the appropriate granularity for phase detection and the time for capturing the phases for unique characterization of a program [15–18]. Beg [5], and Beg and Ibrahim [6] presented machine-learnt models for predicting processor system performance. Their models used a wider range of hardware (processor and memory) parameters than Joseph et al.'s [11] predictive RBF. In Ref. [6] the authors also proposed that the models be used as a tool for computer architecture pedagogy. These models characterized the complete program trace with a single variable – a somewhat limited representation of a program's dynamic behavior. A processor model that incorporated a large and exhaustive set of hardware (microarchitectural) and dynamic features of programs was introduced by Beg [1]; he used basic block frequencies to represent different programs. However, no numerical analysis or evaluation of different block configurations was done.

4.1.3   A Brief Introduction to Artificial Neural Networks

Artificial *neural networks* (NNs) are electronic equivalents of biological brains. The building blocks of NNs are simple processing entities called *neurons*. The neurons are interconnected to generate outputs in a parallel fashion (as compared to the conventional sequential computers). A *feed-forward neural network* (FFNN)[1] is generally composed of layers of neurons: input, hidden, and output. The outputs of each layer only feed the next layer and not any of the previous layers. (Section 4.3 includes the figure of a simplistic NN). The neurons multiply their inputs values

---

[1] The acronyms NN and FFNN are used alternatively in this work.

with their respective *weights*, before passing them through an *activation function* (such as *sigmoid*) to produce the final neuron output. The neuron weights are determined by training the NNs with some known input examples (*training sets*). The weights are iteratively adjusted in such a way that each set of inputs produces output(s) close to the example's pre-known output(s). An iteration of the weight-tuning process is known as an *epoch*. Some known input-output sets (*validation sets*) are used for validating the NN prediction accuracy. The validation sets are not 'shown' to an NN during training [19].

4.1.4 An Overview of Principal Component Analysis

In cases where the dimension of a dataset is large but individual components in the set are redundant (due to high correlation), it is beneficial to reduce the set's dimensions. The notion behind PCA is that a special coordinate system can be used that is dependent on the *cloud* of points for a dataset containing k numeric attributes. This is done by placing the first axis in the direction of largest variance of the points so that variance along that axis is maximized. The second axis is also chosen in such a way that it maximizes the variance along it. The process continues by choosing each axis to maximize its share in remaining variance. Technically speaking, covariance matrices and eigenvectors are used for PCA. A threshold can be set to attain the reduced dimension dataset, for example, by excluding all components that contribute less than 2% (an arbitrary number) variation to the dataset [19]; the reduced vector dimension has many applications, for example, in processor performance modeling – the subject of this paper.

4.2 Experimental Setup and Data Processing

4.2.1 Data Acquisition

The NN model in our research emulates the behavior of a superscalar processor. The data for this research was collected using a superscalar processor simulator called SimpleScalar (*sim-outorder* module) [13]. A diagram representing the processor architecture is shown in Fig. 4.1. We ran 630 simulations using different configurations of *sim-outorder*, in order to collect the processor's execution efficiency data. The efficiency is measured in terms of instructions (completed) per cycle (IPC). The simulator configurations are listed in Table 4.1. The simulations for ten different SPEC CPU2000 integer benchmarks (namely, *bzip, crafy, gap, gcc, gzip, mcf, parser, perlbmk, vortex,* and *vpr*) made use of '*test*' inputs [12]. To reduce the effect of program initialization, we fast-forwarded the simulations by 50 million cycles [13]. We limited each run to 200 million instructions in order to complete all

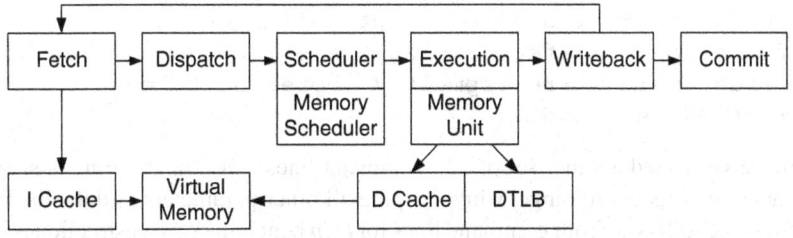

Fig. 4.1 *SimpleScalar's* superscalar architecture (*sim-outorder* module)

Table 4.1 Hardware and software parameter ranges/settings for collecting *SimpleScalar* simulation data

No.	Description	Ranges/values
1	Branch prediction scheme	'Taken,' 'Not-taken,' 'Perfect' (*nominal*)
2	Commit width (no. of instructions)	1, 2, 4, 8, 16, 32, 64
3	Decode width (no. of instructions)	1, 2, 4, 8, 16, 32, 64
4	Fetch queue width (no. of instructions)	2, 4, 8, 16, 32, 64, 128
5	Branch misprediction penalty (cycles)	1, 2, 3, 4, 6, 8, 12, 16, 24, 32, 48, 64, 96, 128
6	Ratio of CPU and bus speeds	2, 4, 8, 16, 32, 64, 128
7	Issue width (no. of instructions)	1, 2, 4, 8, 16, 32, 64
8	Load/store queue (no. of instructions)	2, 4, 8, 16, 32, 64, 128
9	Floating point ALUs	1, 2, 3, 4, 5, 6, 7, 8
10	Floating point multipliers	1, 2, 3, 4, 5, 6, 7, 8
11	Integer ALUs	1, 2, 3, 4, 5, 6, 7, 8
12	Integer multipliers	1, 2, 3, 4, 5, 6, 7, 8
13	Register update unit (no. of instructions)	2, 4, 8, 16, 32, 64, 128
14	Benchmarks (SPEC CPU 2000 integer)	*bzip, crafy, gap, gcc, gzip, mcf, parser, perlbmk, vortex, vpr*

simulations in a reasonable amount of time. The simulations were run on multiple x86-machines running *cygwin* (a UNIX emulator) under Windows-XP. Each of the 630 simulations lasted 1–1.5 h.

First, we created a set of 310 *sim-outorder* command lines by randomly selecting the parameter values listed in Table 4.1. One such command line for *crafty* benchmark is shown below:

```
sim-outorder -fastfwd 50000000 -max:inst 200000000
-cache:il1 il1:64:32:4:t -cache:il2 il2:64:32:16:l
-cache:dl1 dl1:4:8:4:f -cache:dl2 dl2:16:8:4:l
-cache:dl1lat 3 -cache:dl2lat 6 -cache:il1lat 2
-cache:il2lat 8 -tlb:itlb itlb:256:16:2:t -tlb:dtlb
none -tlb:lat 30 -mem:lat 16 2 -mem:width 16
-decode:width 4 -issue:width 8 -commit:width 4
```

```
-ruu:size 16 -lsq:size 8 -fetch:ifqsize 8
-fetch:speed 8 -fetch:mplat 6 -res:ialu 3 -res:imult
7 -res:fpalu 1 -res:fpmult 7 -bpred nottaken
crafty00.peak.ev6
```

Then, we generated another set of 320 command lines, varying at a time, a single
parameter over its entire range, while keeping all other parameters as defaults. For
example, the following four command lines for *bzip* benchmark use instruction issue
unit widths of 1, 2, 4 and 8 (All seven values of issue unit widths are shown in row
7 of Table 4.1):

```
sim-outorder -fastfwd 50000000 -max:inst 200000000
-issue:width 1 bzip200.peak.ev6 input.random 2

sim-outorder -fastfwd 50000000 -max:inst 200000000
-issue:width 2 bzip200.peak.ev6 input.random 2

sim-outorder -fastfwd 50000000 -max:inst 200000000
-issue:width 4 bzip200.peak.ev6 input.random 2

sim-outorder -fastfwd 50000000 -max:inst 200000000
-issue:width 8 bzip200.peak.ev6 input.random 2
```

4.2.2 Dimension Reduction

A goal of this work was to investigate whether the count of hardware and software
related parameters for a processor model could be reduced while keeping the performance prediction accuracy reasonably high. Besides containing hardware-related
data, *sim-outorder*'s text-based log files also included basic block data required for
program characterization. We used a Perl script to collect the basic-block distribution data from *SimpleScalar* run log files. Figure 4.2 shows a histogram of program
basic block sizes 1 to 20, while executing *bzip* benchmark. The blocks larger than
20 instructions are included in the last bin. One can notice that most instructions
appear in very small block sizes, which could be attributed to the repetitive nature

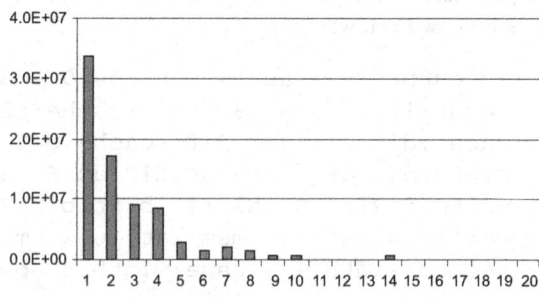

Fig. 4.2 Block distribution
of *bzip* benchmark. Block
sizes 1 to 20 are shown. Any
blocks larger than 20
instructions are included in
the last bin

Fig. 4.3 Block distribution of *crafty* benchmark. Block sizes 1 to 20 are shown. Any blocks larger than 20 instructions are included in the last bin

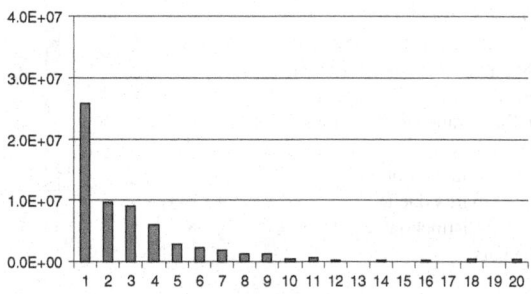

Fig. 4.4 Block histogram for ten different benchmarks for 'Setup 1' (Table 4.2). Block sizes 1–4 are much larger in number than larger ones, causing the grouping to be not too suitable for distinguishing benchmarks

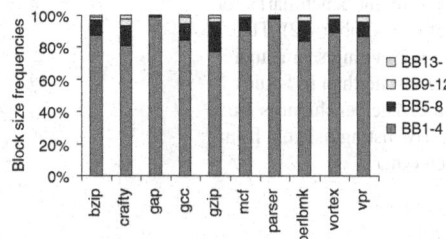

of this compression utility. Similarly, Fig. 4.3 shows the data for *crafty* benchmark; the data is relatively widely distributed as compared to *bzip*.

We made many arbitrary groups of different block sizes and studied how relevant the groups were to the efficiency metric (IPC) of a processor. Figures 4.4–4.7 show basic block histograms for 'Setups 1–4' of Table 4.2. Figures 4.4 and 4.5 indicate that for all ten benchmarks, the first sets (BB1–2 and BB1–4, respectively) are overwhelmingly larger than others, and thus may not be suitable to represent the respective benchmarks in an IPC prediction model. In comparison, 'Setups 3 and 4' plotted in Figs. 4.6 and 4.7, respectively, clearly distinguish the stacked bars for different benchmarks, hence making them better candidates for representing software/benchmarks in the model.

We further analyzed block data, this time in numerical terms. We employed Matlab's `corrcoef` (correlation coefficient/CC) function for studying how well different blocks correlate with IPC. Table 4.2's cells marked with 'x' show the parameters which are highly correlated (>0.5) to IPC.[2] In all five Setups, a common set of microarchitectural parameters appears; only 6 out of 13 hardware parameters

---

[2] Legends for Table 4.2:

CC = correlation coefficient > 0.5; *bPred* = branch predictor type (nominal); *commit* = commit unit width (no. of instructions); *decoder* = decoder unit width (no. of instructions); *fetchIFQ* = instruction fetch unit size (no. of instructions); *brPredLat* = fetch latency (no. of cycles); *busRatio* = ratio of front-end speed to processor speed; *issue* = issue unit width (no. of instructions); LSQ = load/store queue (no. of instructions); *fpALU* = no. of floating point ALUs; *fpMult* = no. of floating point multipliers; *iALU* = no. of integer ALUs; *iMult* = no. of integer multipliers; RUU = capacity of register update units (no. of instructions); *BB1* = no. of basic blocks contain-

Fig. 4.5 Block histogram for ten different benchmarks for 'Setup 2' (Table 4.2). Larger blocks are better represented in this arrangement. Setups 1 and 2 may not good representations of benchmarks due to not-so-distinct nature of the bars

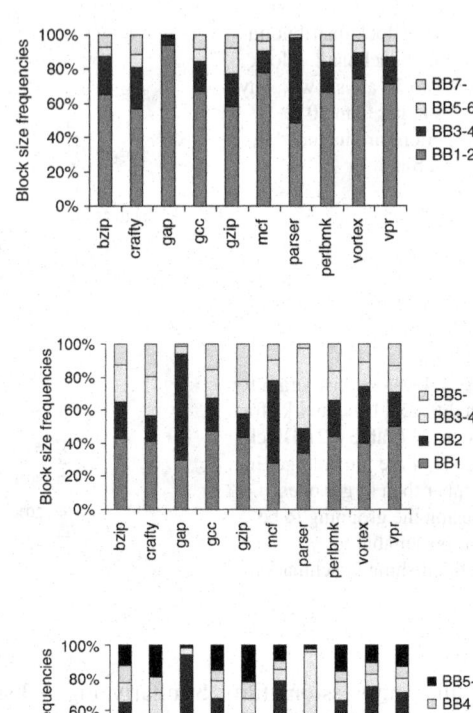

Fig. 4.6 Block histogram for ten different benchmarks for 'Setup 3' (Table 4.2). The block grouping seems more appropriate than in Setups 1 and 2. The benchmarks are clearly distinguishable from each other

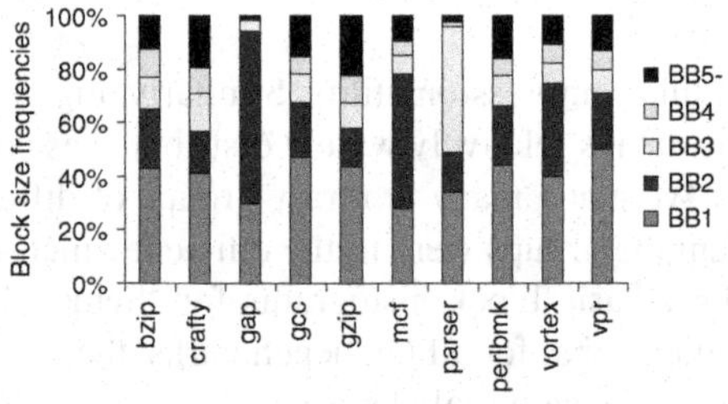

Fig. 4.7 Block histogram for ten different benchmarks for 'Setup 4' (Table 4.2). This grouping of blocks shows the most variety in block distribution thus representing different benchmarks more distinctly than other setups

seem sufficient for an IPC model. In Setup 0, most basic block values are needed for the model – not much of dimension reduction here. Setups 1 and 2 show that only three groups of basic blocks correlate highly with IPC, while Setup 3 has four block groups, and Setup 4 has five groups correlating with IPC. Note that in our experiments, we had used only integer benchmarks; so the presence of floating point units (ALUs and multipliers) did not have any effect on IPC.

Refer to Table 4.2 again. As compared to correlation coefficients, PCAs show the need to use a higher dimension model (and thus were not utilized for NN models). Setup 0 yielded a 16-component set, while all other Setups produced 15-component sets with these Matlab commands:

```
[arrP,transMat] = prepca(pn,0.02); % 2 percent cutoff
[R,C] = size(arrP); % reduced data size
```

---

ing one instruction; *BB1–4* = no. of basic blocks containing one to four instructions, *BB5-* = no. of basic blocks containing five or more instructions, etc.

Table 4.2 Microarchitectural and software parameters and their relationships to the processor throughput. The reduced sets of parameters are marked with x's (legends for this table are included as a footnote on page 3)

	Setup 0		Setup 1	CC	Setup 2	CC	Setup 3	CC	Setup 4	CC
	Parameter		Parameter	CC	Parameter	CC	Parameter	CC	Parameter	CC
Hardware/microarchitecture	bPred	x	bPred	x	bPred	x	bPred	x	bPred	x
	commit		commit		commit		commit		commit	
	decoder	x	decoder	x	decoder	x	decoder	x	decoder	x
	fetchIFQ		fetchIFQ		fetchIFQ		fetchIFQ		fetchIFQ	
	brPred	x	brPred	x	brPred	x	brPred	x	brPred	x
	Lat		Lat		Lat		Lat		Lat	
	busRatio		busRatio		busRatio		busRatio		busRatio	
	issue	x	issue	x	issue	x	issue	x	issue	x
	LSQ	x	LSQ	x	LSQ	x	LSQ	x	LSQ	x
	fpALU		fpALU		fpALU		fpALU		fpALU	
	fpMult		fpMult		fpMult		fpMult		fpMult	
	iALU		iALU		iALU		iALU		iALU	
	iMult		iMult		iMult		iMult		iMult	
	RUU	x	RUU	x	RUU	x	RUU	x	RUU	x
Software (basic blocks)	BB1	x	BB1–4	x	BB1–2	x	BB1	x	BB1	x
	BB2	x	BB5–8	x	BB3–4	x	BB2	x	BB2	x
	BB3	x	BB9–12	x	BB5–6	x	BB3–4	x	BB3	x
	BB4	x	BB13-	x	BB7-	x	BB5-	x	BB4	x
	BB5	x							BB5-	x
	BB6	x								
	BB7	x								
	BB8	x								
	BB9	x								
	BB10	x								
	BB11	x								
	BB12	x								
	BB13	x								
	BB14	x								
	BB15	x								
	BB16									
	BB17									
	BB18									
	BB19	x								
	BB20									
	Total using correlation coeff.	22		9		9		10		11
	Total using PCA	16		15		15		15		15

4.3   Neural Network Structure and Training

Figure 4.8 shows the structure of an FFNN with software and hardware inputs. Each of the NN inputs – not all inputs are shown in the figure – of the NN model corresponds to a single neuron. A single 'continuous' neuron in the output layer represents NN's IPC predictions.

It is well known that an NN with a single hidden layer is able to model most non-linear systems, so we limited our experiments to one-hidden layer NNs. In Table 4.1, we can notice the non-linearity of input parameters, which can adversely affect the learn-ability of a NN, so we applied log_2 to such inputs, as a data *pre-processing* step. We used Matlab's newff() command to create the *back-propagation* FFNNs. NN training set comprised of 50% of full data set. While 25% of all data sets were used for validation, and the remaining 25% datasets for testing. For every NN configuration, we conducted three or more training sessions completely independent of each other, in order to find the best performing NN, and to avoid the possibility of local minima. Figure 4.9 shows training progress for one of the NNs. And Table 4.3 shows NN training, validation and testing statistics for 24 different NN configurations. Given enough epochs, most NN configurations were able to reach better than 95% prediction accuracy. In general, Setup 0 trained faster and better than other Setups. For Setups 0 and 1, four-neuron hidden layers fulfilled the accuracy criteria, although Setup 0 has 22 inputs and Setup 1 just 9. Setup 2 required at least five neurons in hidden layer, while Setup 3 did 7.

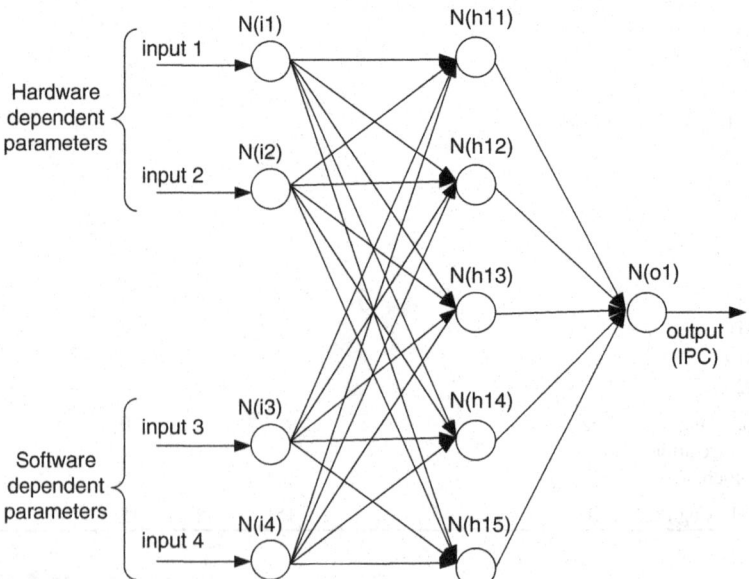

Fig. 4.8 A three-layer feed-forward neural network with hardware and software-related inputs

Fig. 4.9 Training progress
for an NN with Setup 0:
inputs = 22; hidden
neurons = 3; and output
neurons = 1

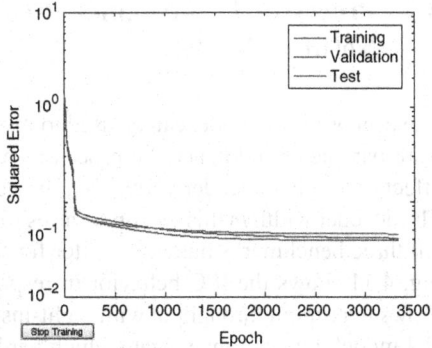

Table 4.3 Neural network training statistics: parameters (Setup nos. 0–3), neurons in different layers, and training performance. Matlab settings: `net.trainParam.goal = 0.01;` `net.trainParam.lr = 0.01; net.trainParam.epochs = 40000;` threshold functions for hidden and output layers = {'tansig','tansig'}; training function = 'traingd'

Set-up no. (Table 4.2)	Neuron count [input hidden output]	Epochs	Training error	Validation error	Testing error	"net. train "Param. goal" met?
0	[22 3 1]	3,334	4.1%	3.9%	4.5%	Yes
0	[22 4 1]	13,165	2.5%	2.9%	4.1%	Yes
0	[22 5 1]	12,349	2.5%	2.6%	3.9%	Yes
0	[22 7 1]	11,884	2.5%	2.6%	3.3%	Yes
0	[22 9 1]	7,542	2.5%	2.8%	3.2%	Yes
0	[22 12 1]	11,065	2.5%	3.1%	3.2%	Yes
1	[9 3 1]	40,001	2.8%	3.5%	3.7%	No
1	[9 4 1]	31,693	2.5%	3.2%	4.6%	Yes
1	[9 5 1]	26,150	2.5%	2.7%	3.2%	Yes
1	[9 7 1]	14,093	2.5%	2.8%	3.9%	Yes
1	[9 1]	17,824	2.5%	3.1%	3.8%	Yes
1	[9 12 1]	12,487	2.5%	3.2%	4.5%	Yes
2	[9 3 1]	40,001	3.6%	4.1%	4.8%	No
2	[9 4 1]	38,216	2.5%	3.3%	3.5%	No
2	[9 5 1]	34,522	2.5%	3.5%	3.3%	Yes
2	[9 7 1]	26,167	2.5%	3.4%	4.0%	Yes
2	[9 9 1]	27,406	2.5%	2.8%	4.0%	Yes
2	[9 12 1]	20,860	2.5%	2.7%	3.5%	Yes
3	[10 3 1]	33,729	2.5%	3.8%	35.6%	No
3	[10 4 1]	31,238	2.5%	3.1%	4.4%	No
3	[10 5 1]	11,897	3.4%	3.6%	5.0%	No
3	[10 7 1]	18,541	2.5%	3.2%	3.5%	Yes
3	[10 9 1]	10,743	3.1%	3.6%	4.7%	Yes
3	[10 12 1]	17,779	2.5%	2.9%	3.5%	Yes

4.4 Processor Performance Prediction with the Neural Network Model

The proposed NN model can be used to investigate how different hardware and software parameters influence the processor performance. Firstly, Fig. 4.10 shows the effect of varying decoder width, i.e., the number of instructions decoded per cycle. The decoder width in this example varies from 1 to 64 for three different benchmarks All three benchmarks flatten out after fourinstruction wide decoder unit. Secondly, Fig. 4.11 shows the IPC behavior in response to commit unit variations. The unit shows maximum throughput with eight-instruction width. With other runs using the NN model, one can investigate which hardware elements (ALU, issue width, etc.) or the software characteristics (parallelism in a program, branch frequencies, etc.) are the limiting factors in processor performance.

To illustrate how the dynamic nature of a program affects IPC, we created two artificial program traces, T1 and T2 (Fig. 4.12). T1 has small percentages of 1 and 2-instruction blocks and much larger percentage of larger (five or more instructions) blocks. Such a trace may come from programs in which loop unrolling has been done. T2's distribution is just the opposite of T1, i.e., there are more blocks of smaller sizes (1 or 2 instructions) and the smaller number of blocks are of larger sizes (five or more instructions). When the block frequencies of T1 and T2 are input

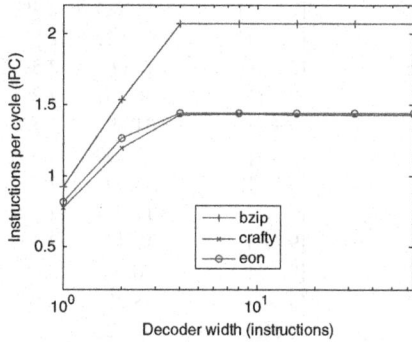

Fig. 4.10 Sensitivity of IPC to decoder width, while all other parameters are set to *sim-outorder* defaults. Only four-instruction wide decoders are able to produce the best IPC

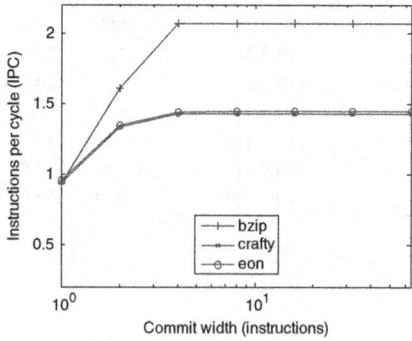

Fig. 4.11 Sensitivity of IPC to commit width, while all other parameters are set to *sim-outorder* defaults. Only a four instruction-wide commit unit is sufficient under the given conditions

Fig. 4.12 Traces with different block distributions exhibit different IPCs, while all hardware parameters are set to *sim-outorder* defaults. IPC for T1 is 1.15 and IPC for T2 is 1.30

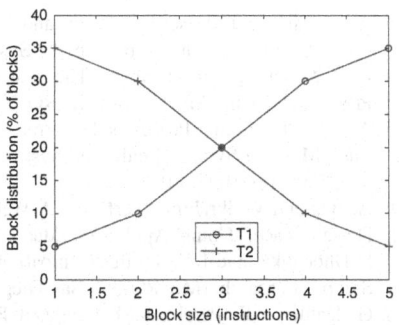

to the NN model, we observe that T1 has 33% lower IPC than T2; this may be due to the small cache (number of sets and line size) used in the experiment; smaller caches can cause thrashing and thus lower the execution throughput. On the other hand, T2's smaller blocks may have better chances of fitting in cache lines making the fetch stage efficient and thus improving the overall execution throughput.

A point worth noting: the examples presented above, if ran on a traditional detailed, cycle-accurate simulator may have required several computer-days. Whereas, the NN model produces the same results almost instantly. The model's speed efficiency can be helpful for researchers and students alike.

4.5   Conclusions

Using the NN model proposed in this research, one can estimate the performance of a processor without requiring lengthy simulations. Instead of using an exhaustive set of microarchitectural features of a processor and the program characteristics, the model is based on a reduced set of parameters. A designer can quickly make educated guesses about the micro-architectural parameters for optimum hardware performance, while a compiler designer can speedily observe the effects of code characteristics on the overall execution throughput. The model can also find applications in the teaching of computer architecture and compiler design. As a continuation of the current research, we are looking into: (1) developing an analytical model that includes both hardware and software related parameters for prediction of a (uni-)processor's performance, and (2) creating a multi-processor performance prediction model.

References

1. A. Beg, "Incorporating program characteristics into a processor model," International Conference on Machine Learning and Data Analysis 2008 (ICMLDA'08), 2008.
2. S. Wallace and N. Bagherzadeh, "Modeled and Measured Instruction Fetching Performance for Superscalar Microprocessors," IEEE Trans. Parallel Distrib. Syst., Vol. 9, No. 6, Jun. 1998, pp. 570–578.

3. A. Hossain, D. J. Pease, J. S. Burns, and N. Parveen, "A Mathematical Model of Trace Cache," Proc. IEEE Inter. Conf. Appl. Specific Syst., Archit. (ASAP'02), San Jose, CA, July 2002.
4. D. B. Noonburg and J. P. Shen, "Theoretical Modeling of Superscalar Processor Performance," Proc. Inter. Symp. Microarch. (MICRO-27), San Jose, CA, Nov. 1994.
5. A. Beg, "Predicting Processor Performance with a Machine Learnt Model," Proc. 50th IEEE Inter. Midwest Symp. Circuits and Syst., (MWSCAS/NEWCAS'07), Montreal, Canada, Aug. 5–8, 2007, pp. 1098–1101.
6. A. Beg and W. Ibrahim, "PerfPred: A Web-Based Tool for Exploring Computer Architecture Design Space," Comp. Appl. Eng. Educ., vol. 17, issue 3, pp. 305–313, 2009.
7. S. Dhodapkar and J. E. Smith, "Comparing Program Phase Detection Techniques," Proc. Inter. Symp. Microarch. (MICRO-36), San Diego, CA, Dec. 2003.
8. G. Hamerly, E. Perelman, J. Lau, and B. Calder, "Simpoint 3.0: Faster and More Flexible Program Analysis," J. Instr. Level Parallelism (JILP), (website) http://www.cse.ucsd.edu/~calder/papers/JILP-05-SimPoint3.pdf, 2005.
9. SPEC CPU95 benchmarks, (website) http://www.spec.org/cpu95/.
10. R. E. Wunderlich, T. F. Wenisch, B. Falsafi, and J. C. Hoe, "SMARTS: Accelerating Microarchitecture Simulation via Rigorous Statistical Sampling," Proc. Inter. Symp. Comp. Arch. (ISCA 2003), San Francisco, CA, June 2003.
11. P. J. Joseph, K. Vaswani, and M. J. Thazhuthaveetil, "A Predictive Performance Model for Superscalar Processors," Proc. IEEE/ACM Inter. Symp. Microarch. (MICRO'06), Orlando, FL, Dec. 2006.
12. SPEC CPU2000 benchmarks, (website) http://www.spec.org/cpu2000/.
13. SimpleScalar LLC, (website) http://www.simplescalar.com.
14. W. Lee, K. Patel, and M. Pedram, "B2Sim: A Fast Micro-architecture Simulator Based on Basic Block Characterization", Proc. IEEE/ACM/IFIP Conf. Hardware/Software Codesign and Syst. Synthesis (CODES + ISSS'06), Seoul, South Korea, Oct. 2006, pp. 199–204.
15. T. Sherwood and B. Calder, "Time Varying Behavior of Programs," UCSD Technical Reports, (website) http://www-cse.ucsd.edu/~calder/papers/UCSD-CS99-630.pdf, Aug. 1999.
16. T. Sherwood, E. Perelman, and B. Calder, "Basic Block Distribution Analysis to Find Periodic Behaviors and Simulation Points in Applications," Proc. Inter. Conf. Parallel Microarch. and Compilation Techniques (PACT-2001), Barcelona, Spain, Sept. 2001, pp. 3–14.
17. T. Sherwood, S. Sair, and B. Calder., "Phase Tracking and Prediction," Proc. Inter. Symp. Comp. Arch. (ISCA 2003), San Francisco, CA, June 2003, pp. 336–347.
18. A. S. Dhodapkar and J. E. Smith, "Comparing Program Phase Detection Techniques," Proc. Inter. Symp. Micro-arch. (MICRO-36), San Diego, CA, Dec. 2003. pp. 217–227.
19. I. Witten and E. Frank. Data Mining. Morgan Kaufmann, NY, 2005.

Chapter 5
Hybrid Machine Learning Model
for Continuous Microarray Time Series

Sio-Iong Ao

Abstract A hybrid machine learning model of the principal component analysis and neural network is described for the continuous microarray gene expression time series. The methodology can model numerically the continuous gene expression time series. The proposed model can give us the extracted features from the gene expressions time series with higher prediction accuracies. It can help practitioners to gain a better understanding of a cell cycle, and to find the dependency of genes, which is useful for drug discoveries. In this chapter, we describe the background, the machine learning algorithms, and then the application of the hybrid machine learning in the microarray analysis. The machine learning model is compared with other popular continuous prediction methods. Based on the results of two public microarray datasets, the hybrid method outperforms the other continuous prediction methods.

Keywords Machine learning · Neural network · Microarray · Gene expression · Time series prediction

5.1 Introduction

With the advances in DNA microarray technology [10], the gene expression values at different time points of a cell cycle can be obtained. In the simplest case, two time points are taken: before and after an event. A more comprehensive study will involve the taking of values at different periods. The frequencies of the time points can have ranges from several minutes to several hours. Different methods have been developed to analyze gene expression time series data, see for instance [15, 16, 20, 27, 31, 34, 41, 45–47, 55, 58, 60 etc.]. It is noted that their approaches are different from our hybrid principal component analysis (PCA) and neural network (NN)

S.-I. Ao (✉)
Harvard School of Engineering and Applied Sciences, Harvard University, Cambridge, MA, USA
e-mail: siao@comlab.ox.ac.uk; siao@harvard.edu

S.-I. Ao et al. (eds.), *Advances in Machine Learning and Data Analysis*,
Lecture Notes in Electrical Engineering 48, DOI 10.1007/978-90-481-3177-8_5,
© Springer Science+Business Media B.V. 2010

model for gene expression time series. About half of these researches focused on building special clustering algorithms for these time series data, while the others tackled the problems of inferring systems of linear differential equations, the visualizing of the gene data and the determination of their periodicity. Instead, with the PCA-NN algorithm [3,4], it has the advantage of the nonlinear flexibility of the neural network. The AIC test is employed to determine the optimal lag length used in our models, whereas, in the existing models, only one lag length (or a fixed number in a few cases) for each gene expression value change is included. The lag length refers to the number of lags (the number of previous values of the variable) used in the model.

5.1.1 Computational Methods for Microarray Time Series Analysis

Various methods like self-organizing maps [37], k-nearest neighbor [2] and hidden Markov models [28] have been employed for the microarray analysis. These studies mainly focus on the clustering and the measurement of the similarity among the different expressions. For the gene expression time series analysis, methods like warping algorithms [1], the comparison of similarity functions of the genes [9], the identification of gene regulatory networks with graph method [11], and dynamic models [17] etc., have been developed.

Special clustering algorithms were employed to explore the gene expression time series data from the microarray experiments. Costa et al. [15] have proposed the symbolical description of multiple gene expression time series. Each variable will take as a set of values in a time series and the results are compared with Self-Organizing Map algorithm. Yoshioka and Ishii [58] have employed a clustering method based on mixture of constrained PCA. It can classify genes with similar expression patterns into the same cluster regardless of their magnitude (scale). In the study [46], Tabus and Astola have handled the problem of the non-uniformly sampling of the gene expression time series. The minimum description length model is fitted to each gene and then the optimum parameters are used for clustering the genes. The extrapolation of the gene expression time series data by the minimum description length model can be applied in our methodology too for non-uniformly sampling data.

Syeda-Mahmood [45] studied a clustering algorithm that uses the scale-space distance as a similarity metric. The scale-space analysis is to detect the sharp twists and turns of the gene time series and to form the similarity measure between time profiles. Wu et al. [55] have developed a procedure for the determination of the minimal number of samples or trials required in a microarray experiment for clustering. The procedure is an incremental process that will terminate when the evaluation of the results of two consecutive experiments of k-means clustering shows they are sufficiently close. Jiang et al. [27] use a density-based approach to identify the clusters such that the clustering results are of high quality and robustness. Futschik and Kasabov [20] employ the fuzzy c-mean (FCM) clustering to achieve a robust analysis of gene expression time series. The issues of parameter selection and cluster validity are also addressed.

The construction of genetic network from gene expression time series was tackled in Refs. [31, 41, 47]. Kesseli et al. have employed monotonic time transformations (MTT) for inferring a Boolean network. Several different methods of clustering have been used to form different transformations. Tabus et al. build systems of differential equations for specifying the genetic networks. The structure of the networks is inferred by operating with the exact solutions of the linear differential equations, which are obtained through the eigenvalue decomposition of the system matrix. Sakamoto and Iba have also used a system of ordinary differential equations as a model of the network and infer their right-hand sides by using genetic programming (GP) instead. The least mean square (LMS) method is used along with the GP to explore the search space more effectively in the course of evolution. In these systems of linear differential equations, there is a strong assumption that the genetic interactions are linear.

The visualizing of the gene expression time series is discussed in studies [16, 60]. Zhang et al. have introduced the first Fourier harmonic projection (FFHP) to translate the multi-dimensional time series data into a two-dimensional scatter plot. The spatial relationship of the points reflects the structure of the original dataset and the relationships among clusters become two-dimensional. Craig et al. propose the display technique that operates over a continuous temporal subset of the time series, with direct manipulation of the parameters defining the subset. Its advantage is that the number of elements being displayed will not be reduced.

Langmead et al. [34] formulate the task of estimating an expression profile's periodicity and phase as a simultaneous bicriterion optimization problem. The maximum entropy-based analysis technique is employed for extracting and charactering rhythmic expression profiles, and is found to work better than the Fourier-based spectral analysis for signals in the microarray experiments. Yeang and Jaakkola [56] explain time correlations between gene expression profiles through factor-gene binding information to estimate latencies for transcription activation. This can estimate latencies for transcription activation. The resulting aligned expression profiles are subsequently clustered and again combined with binding information to determine groups or subgroups of co-regulated genes.

5.2 Machine Learning Methods

The machine learning methods have often been employed for pattern matching and pattern discovery. Machine learning is itself a collection of methods that result from the convergence of several disciplines like statistics, biological modeling, adaptive control theory and artificial intelligence (AI). Its spectrum of tools is wide and includes inductive logic programming, genetic algorithms, neural network, Bayesian networks, and hidden Markov Models, etc. Regardless of the divergences of the underlying technology, we can see that they usually have the following steps [7]. First, the input data are fed into a comparison engine that compares the data with assumed model. Then, the comparison results are used for initializing some changes to the data or some modifications to the assumed model. The evaluation engine gives us the performance results based on the modified model. These results are checked against our pre-defined criterion. If the criteria are not met, the above steps are repeated until the stopping criteria are satisfied. In this chapter, the neural network is employed in the hybrid method and is described in the following section.

5.2.1 Neural Network Models

Among the different tasks that machine learning tools can handle, one popular task is to filter the noises of the source data and then to made prediction basing on the extracted patterns of the source data. The neural network has been found to perform this filtering and prediction capability well. The network can first extract the vital signals and information from the source data. Then it can predict what the future signals and information will be, based on some function approximation. The filtering and prediction capability of the neural network have enabled it to become a popular advance tool for the time series prediction, e.g., the financial prediction of the future index, stock share prices, currency rates, the weather forecast, the traffic control forecast and the medical analysis.

The study of the neural network began after the Warren McCulloch and Walter Pitts have proposed the first mathematical model of a single idealized biological neuron in 1943 [36]. The model has been known as McCulloch-Pitts model, which consists of a single neuron that receives the input signals and sums them up with different weights. Then, newer models like the Perceptron by Frank Rosenblatt [40] and the ADALINE by Widrow [53] were developed.

Since these earliest works on the neural network, there have come many other neural network models that made use of the neuron concept. The network models that utilize more than one neuron and contain no feed-back paths within the network are given the term feedforward networks. In the feedforward network, there are the single-layer feedforward networks, which consist of the input layer and output layer only, and the multi-layer feedforward networks, which consist of the input layer, hidden layer and the output layer. In our research, we have utilized the multi-layer feedforward network for building our hybrid models for continuous microarray time series analysis.

5.2.2 Neural Network Models for Microarray Analysis

The function approximation capability of the neural network is one of the network's major properties and advantages. With this property, the researchers can be assured that, provided that appropriate network structure has been employed, the neural network can approximate the real problem accurately. The objective of the function approximate can be formulated as finding function $\Im(.)$ such that:

$$\left\| \Im(x^k) - f(x^k) \right\| < \varepsilon$$

where ε is a small positive number, a set of N different input points are denoted as $\{x^k \in \Re^p, k = 1, 2, \ldots, N\}$, a set of N output points $\{d^k \in \Re^q, k = 1, 2, \ldots, N\}$, and the actual nonlinear input–output mapping between x and d is denoted as: $f(x^k) = d^k$, where the function $f(.)$ is unknown. The mapping function $\Im : \Re^p \to \Re^q$ should closely fulfills the relationship: $\Im(x^k) = d^k$, for k = 1, 2, ..., N. The property of the function approximation by the neural network is clear with the following theorem of the Universal Approximation Theorem of the network.

Definition 5.1. The Universal Approximation Property is said to be satisfied when, with an appropriate number of neurons and optimal weight vector, a neural network can approximate any continuous function, on any compact subset $C \subset \Re^n$ of the input space, to an arbitrary level of accuracy.

Note 5.1. A set $S \subset \Re^n$ is called compact if it is closed and bound. A set is closed if and only if its complement in $\Re^n$ is open. A set $S \subset \Re^n$ is open if for every vector $x \in S$, there is an ε-neighbourhood of $x : N(x, \varepsilon) = \{z \in \Re^n \| \|z - x\| < \varepsilon\}$, such that $N(x, \varepsilon) \in S$. A set is bounded if there is $r > 0$ such that $\|x\| < r$ for all $x \in S$.

Universal Approximation Theorem [21]: Let $\varphi(.)$ denote a bounded and monotone-increasing continuous function. Use I_p to denote the p-dimensional unit hypercube $[0, 1]^p$. Let $C(I_p)$ be the space of continuous functions on I_p. For any function $f \in C(I_p)$ and $\varepsilon > 0$, there exists an approximate function $\Im(.)$ that satisfies:

$$\left| \Im(x_1, \ldots, x_p) - f(x_1, \ldots, x_p) \right| < \varepsilon$$

for all $\{x_1, \ldots, x_p\} \in I_p$, where $\Im(.)$ is defined as:

$$\Im(x_1, \ldots, x_p) = \sum_{i=1}^{M} \alpha_i \varphi \left(\sum_{j=1}^{p} w_{ij} x_j - \theta_i \right)$$

for $i = 1, \ldots, M$ and $j = 1, \ldots, p$, where M is an integer, and α_i, θ_i, and ω_{ij} are sets of real constants.

Barron [5, 6] has estimated that the mean integrated squared error between the target function $f(.)$ and the estimated function $\Im(.)$ is bounded by:

$$O \left(\frac{C_f^2}{M} \right) + O \left(\frac{M_p}{N} \log N \right)$$

where C_f is the first absolute moment of the Fourier magnitude distribution of the target function $f(.)$, M is the total number of hidden nodes, p is the number of input nodes, and N is the number of training sets.

The multi-layer feedforward network is a member of supervised learning. In supervised learning, a training set (represented by an input vector x) and the corresponding desired output vector d are presented to the network. The network is then trained to learn how to minimize the error between its actual output vector z and this desire output vector d [25].

With above function approximation capability, it is not a surprise that the neural network has been reported for its successful applications in the gene expression analysis. Herrero et al. [23] have applied neural network for clustering gene expression patterns. Peterson and Ringner [38] have analyzed tumor gene expression profiles with the network. Sawa and Ohno-Machado [42] have developed a neural network-based similarity index for clustering DNA microarray data. Predictions of TP53 gene sequence of values A, C, G and T have been modeled with neural network in study [44]. Veiga et al. [51] applied the neural network to E. coli for predicting the transcriptional regulatory interactions. Lynn et al. [35] built a neural network for common complex disease analysis.

5.2.3 Dimension Reduction and Transformation

Before extracting meaningful patterns from the data, there are sometimes the needs for the transformation and reduction of the data. These needs may arise due to the fact that the original dataset is too large in dimension. Both reduction and transformation can support the data-mining process when used properly. The datasets may be reduced to the minimum possible size by tactics like sampling or summary statistics etc., while still satisfying our analysis requirement. The transformation can be achieved by translating one type of data to another through mathematical operations or mappings. The noise level in the data may be reduced by the eliminating irrelevant components with suitable transformation. Transformation tools like principal component analysis (PCA) and independent component analysis (ICA) can be employed to find out the dominant components of the dataset. We have applied these algorithms for the microarray time series data, and we will compare their respective performances in our hybrid models.

5.2.4 Principal Component Analysis

Among the tools of the dimension reduction and transformation, the principal component analysis (PCA) is a popular tool for many researchers. Its basic idea is to find the directions in the multidimensional vector space that contribute most to the variability of the data. The representation of data by the PCA consists of projecting the data onto the k-dimensional subspace according to

$$x' = F(x) = A^t x$$

where x' is the vectors in the projected space, A^t is the transformation matrix which is formed by the k largest eigenvectors of the data matrix, x is the input data matrix. Let $\{x_1, x_2, \ldots, x_n\}$ be the n samples of the input matrix x. Then, the principal components and the transformation matrix can be obtained by minimizing the following sum of squared error:

$$J_k(a, x') = \sum_{h=1}^{n} \left\| \left(m + \sum_{i=1}^{k} a_{hi} x_i' \right) - x_h \right\|^2$$

where m is the sample mean, x_i' the i-th largest eigenvector of the co-variance matrix, and a_{hi} the projection of x_h to x_i'.

The principal component analysis was applied to reduce the dimensionality of the gene expression data in studies [8, 24, 49, 57, etc.]. The focuses are on the effective dimensional reduction by the PCA, the analysis of the compressed space and the assistance of the PCA for the classification and the clustering. For example, Hornquist et al. has concentrated on the determination of the effective PCA dimensionality. Bicciato et al. has described how to use the PCA to reduce the gene expression's dimensional base for a better understanding of its basic biology and to have a better classification result. Taylor et al. have applied the PCA to help understand the basis of plant genotype discrimination. Yeung and Ruzzo have tested the efficiency of the PCA for clustering gene expression data. Chen et al. [12] proposed a supervised PCA for gene set enrichment of microarray data.

Khan et al. [32] have applied the PCA and neural network for the classification of cancers using gene expression profiling. Khan's purpose is on the classification and the neural network is trained as a classifier for discrete outputs EWS, RMS, BL, and NB of the cancer types. The PCA has been employed for the dimensionality reduction of the samples. This can avoid the "over-training" of the network (i.e. low number of parameters as compared to the number of samples). Similarly, we have applied the PCA method in our modeling for dimensionality reduction and avoidance of over-fitting.

5.2.5 Independent Component Analysis

The independent component analysis (ICA) is a recently developed theory ([14, 26] and [29]). Its objective is to make the transformed entries mutually independent [50]. Mathematically, let the input samples denoted by x. The task is to determine an N-by-N invertible matrix W, so that the entries $y(i)$, for $i = 0, 1, \ldots, N-1$, of the transformed vector: $y = Wx$, are mutually independent. Statistically, the requirement for the independence is a stronger condition that the condition of the PCA, which only requires the un-correlation of the components. For Gaussian random variables, these two conditions are then equivalent to each other.

The original motivation for the development of the ICA is as followed: Assume that the input data vector x is indeed from a linear combination of statistically independent components. An example is that, of several woofers located in different positions of a room, we have some detectors for checking the sound signals and then for determining the sources from these observed signals. Formally, we have $x = Ay$, where A is known as the mixing matrix of the components y. The task is to determine the de-mixing matrix W as the above paragraph, so that $y = Wx$, for recovering the components of the sources y.

Mathematically, the ICA transformation can work only for non-Gaussian processes, as it is ill-posed for Gaussian processes [50]. Let the independent components y(i) be all Gaussian, then it can be observed that a linear transformation of these components by any unitary matrix will also satisfy the requirement. Thus, PCA should be used in this case, as PCA can return a unique solution by setting a specific orthogonal structure in the transformation. Another condition for the proper working of ICA is that the mixing matrix A must be invertible. For cases where A is a non-square l-by-N matrix, it is required that l must be larger than N and A has to be of full column rank.

The de-mixing matrix W can be estimated by minimizing the mutual information between the transformed random variables. Define the associated entropy of $y(i)$, $H(y(i))$, as (Papoulis, 91):

$$H(y(i)) = -\int p_i(y(i)) \ln p_i(y(i)) dy(i)$$

where $p_i(y(i))$ is the marginal probability distribution function of $y(i)$. Then the mutual information $I(y)$ can be obtained as:

$$I(y) = -H(x) - \ln|\det(W)| + \sum_{i=0}^{N-1} \int p_i(y(i)) \ln p_i(y(i)) dy(i)$$

Techniques of the approximation of this mutual information $I(y)$ and then subsequent minimization of the approximate function can be used for finding the solution of $I(y)$ respective to W [22, 26]. Studies like Kong et al. [33] applied the ICA for the microarray analysis.

5.3   The Proposed Hybrid PCA-NN Machine Learning Model

The motivation for the hybrid method can be explained by looking at the information transfer mechanism among DNAs, mRNAs and proteins. Genes are used as templates for DNA synthesis. In the transcription process, genes are converted into the messenger RNA (mRNA). While the mRNA is subsequently translated to form proteins, some particular proteins can in turn regulate gene expression profiles. In other words, there exists a complex relationship between the current and future gene expressions values and their lags through this mechanism. This work is to model this complex time series relationship by using a continuous numerical model.

The first step of our PCA-NN system is to form the input vectors for the time series analysis. They are the expression levels of the time points in the previous stages of the cell cycle. Then, these input vectors are processed by the PCA. Thirdly, these post-processed vectors were fed to the neural network predictors. Their outputs are compared in the section of result comparison and we can see that the PCA-NN is the most suitable one among the methods we have compared. With the hybrid method, the influence of each gene on the principal components can be obtained. With the knowledge of the disease gene, we can apply our algorithm to find out the influential genes in the development of such disease genes. Therefore, the suitable enhancing or inhibiting of the expression of these leading genes could lead to more effective control for the disease gene growth. This can certainly help us to gain a better understanding of genes in a cell cycle, for example, which gene can be understood best for its future analysis. The method can be regarded as a nonlinear generalization of the linear models in the previous studies. The PCA is employed for feature extraction and the system diagram was shown in the following figure (Fig. 5.1).

As said, the neural network is one of the machine learning tools that can reduce noises and make prediction reliably. A key property of the neural network is its ability of learning for further improving its performance [25]. The learning process starts with the stimulation by the environment. Then, the neural network will have changes in its structure and parameters as a result of this stimulation. These changes will bring the network improvement in its response to the environment.

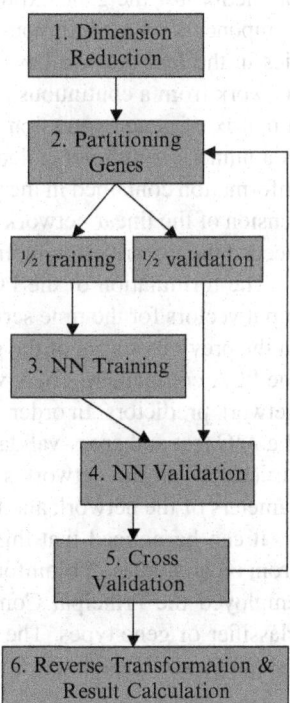

Fig. 5.1 System diagram of the hybrid PCA-NN machine learning model

In the learning process, there can be different objective tasks to achieve, like the function approximation, control, pattern recognition, filtering and prediction etc. The neural network is employed in the hybrid model for the function approximation and prediction of the cell cycles time series microarray data. One advantage of using neural network is that it can give us continuous modeling output. Khan has used the classification function of the network. This work can be viewed as an extension of the discrete models to a continuous modeling of the gene expression. The continuous models have the advantage of resembling the real phenomenon better, as the more intricate aspects of gene regulation are dependent not on whether a gene is transcribed but rather on the level of transcription [54]. In the reverse engineering of the gene profiles, D'haeseieer et al. [18] have studied the linear model for the network inference that can give us the prediction of the genes' development. The model can be written as:

$$x_i(t+1) = \sum_{j=1}^{J} w_{ji} x_j(t) + b_i$$

where x_i is the expression level of the ith gene, J is the number of gene studied. As pointed out in his research, it is possible to further enhance this model with the neural network. And our modeling can be regarded as a nonlinear generalization of this linear model.

Different genetics studies have successfully employed the PCA-NN as a classifier of gene types, with continuous inputs and discrete outputs. In this chapter, the prediction for the gene expression time series are achieved with the PCA and NN components on a continuous numerical inputs and outputs basis. The contribution lies in the fact that we have been developing a more realistic model for the gene network from a continuous prospective. A microarray dataset can be considered as a matrix of gene expression values at various conditions. Each entry in the matrix is a numerical number called expression value. The algorithm can fully utilize the information contained in the gene expression datasets. It can be considered as an extension of the linear network inference modeling, while previous models have often needed the linearity assumption or employed discrete values instead.

The formulation of the PCA-NN model is quite computationally efficient. The input vectors for the time series analysis are the expression levels of the time points in the previous stages of the genes' life cycle. These input vectors are processed by the PCA component. Then, we use these post-processed vectors to feed the neural network predictors. In order to avoid over-training of the network, we have adopted the AIC test and cross-validation to study the optimal setting of the neural network structures and the network's stability. The AIC test can restrict the number of parameters of the network and thus can increase the computational performance.

It can be noticed that this approach for the microarray time series is different from other studies in bioinformatics, like Khan et al.'s [32], which has successfully employed the Principal Component Analysis – Neural Network (PCA-NN) as a classifier of gene types. The first step of our proposed PCA-NN system is to form the input vectors for the time series analysis. They are the expression values of the time points in the previous stages of a cell cycle. Then, these input vectors are processed by the PCA. Thirdly, we use these post-processed vectors to feed the neural network predictors. To our best knowledge, our paper is the first attempt to employ the principal component analysis and neural network to model the gene expression with continuous input and output values.

5.4 The Microarray Time Series Datasets

The first dataset is from the experiment of Spellman et al. [43]. It contains the yeast's gene expression levels at different time points of the cell cycle (18 data points in one cell cycle). From the Spellman's data set, there are totally 613 genes that do not have missing values and that show positive cell cycle regulation by periodicity and correlation algorithms. While the number of variables is large, the number of observations per variable is small (18 time points for each gene).

The average absolute percentage change of the genes between two adjacent time points is 94.92%. We will see later that this large volatility of expression levels makes prediction difficult. This value can be proven to be equivalent to the average percentage prediction error of the Naïve method. The Naïve method is one of the most basic while popular methods for time series analysis, and it simply uses the previous realized gene expression value to predict the next coming value. The underlying assumption is that trends and turning points cannot be predicted, and thus, the horizontal line extrapolation is used as the forecast. In other words, the method is equivalent to a random walk model, which shares the same assumption about the structure of the time series data.

The dataset of Cho et al. [13] was also tested for performance comparison. There are 17 time points for a total of 384 genes in this data set. The prediction error for the Naïve method is 28.52%. A potential difficulty with applying the regression-based methods to the microarray data is the possible non-uniformly sampling of the time series data. This can be solved, for example, by regressing the gene expression levels against the various non-uniform time points. Then, extrapolated uniformly sampled time points can be obtained from the derived regression model. Details for the extrapolation method can be found in Ref. [59] etc.

5.5 Experimental Results

5.5.1 Models with Stand-Alone Neural Network

While the gene activities are highly complicated and nonlinear, the neural network is known for its non-linear capability. In this problem, it is used to check if this non-linear method can provide more accurate numerical forecasting. The typical three-layer neural network architecture is employed. The layers are the input layer, the hidden layer and the output layer. The inspiring idea for this structure is to mimic the working of our brain. The above layers correspond to the axons for inputs, synapses, soma, and axons for outputs.

In the computational experiment with the stand-alone neural network, the inputs x_i's will be the expression levels of the lags (of various length) of each gene in turn. And they will make the current expression level (denoted by y) prediction for the gene. Mathematically, these inputs x_i's are fed into the neural network structure with the output y as followed [39]:

$$y = g \left(\sum_{j=1}^{J} w_j^{(2)} f \left(\sum_{i=1}^{I} w_{ji}^{(1)} x_i \right) \right)$$

where I denotes the number of inputs, J the number of hidden neurons, x_i the ith input, $w^{(1)}$ the weights between the input and hidden layers, $w^{(2)}$ the weights between the hidden and output layers.

The lag length of the input variable will be determined by the AIC method, as we will see later. In the PCA-NN experiment, the x_i's will denote the lags of each principal component value in turn. In our study, we have used the tansig activation function:

$$f(x) = \frac{2}{1 + e^{-2x}} - 1$$

This is mathematically equivalent to the tanh(x), but runs faster in Matlab than the implementation of the tanh [52]. And, we have simply used the linear combination of the inputs for the output activation function.

The results with the neural network show that the prediction is better than other methods compared but the errors are still high. It may be due to the lack of enough training data, and also due to the fact that the gene expression levels are changing so rapidly that the accurate forecasting is difficult to achieve. The sum of the absolute errors of the prediction is 2,340 while the sum of the absolute gene expression values is 3,921 for the dataset 1. The absolute percentage error is found to be 59.68% for a neural network of ten hidden neurons and a lag length of three previous values. This is still a large percentage error when compared with other time series prediction by the neural network, like the short-term financial time series forecasting. The absolute percentage error for the second dataset is 14.76%. We have studied the effects of using different network architectures and of changing the number of feeding terms for the network. To avoid over-training, we have adopted the AIC tests and cross-validation procedure.

5.5.2 Hybrid Algorithms of Principal Component and Neural Network

As the prediction errors made by the stand-alone neural network are still large, the PCA is tested to see if it can assist the neural network for making more accurate prediction. As said above, the PCA has been used successfully in the gene expression data analysis to reduce the dimensionality of the data set for better classification results.

The PCA was applied to the whole spectrum of the genes in the two datasets separately. With the principal components obtained by Singular Value Decomposition (SVD) method, the gene expression matrix of dimension 613×18 has been reduced to 17×18. In the following Figs. 5.2 and 5.3, the first, second and third principal components are shown. We can see that there exist some more or less clear trends for the neural network to make the prediction. The other major components have similar property too. These principal components will serve as the input vectors for the neural networks. The purpose of dimension reduction can be said successful.

Then, the role by the neural network in the assistance of prediction performance was evaluated. The neural network was used to make prediction for each principal component. Then, the predictions are transformed back into the original vector space. In the computational experiment with Spellman's dataset, the total absolute prediction error of the neural network in the training (with values at t-1, t-2 and t-3

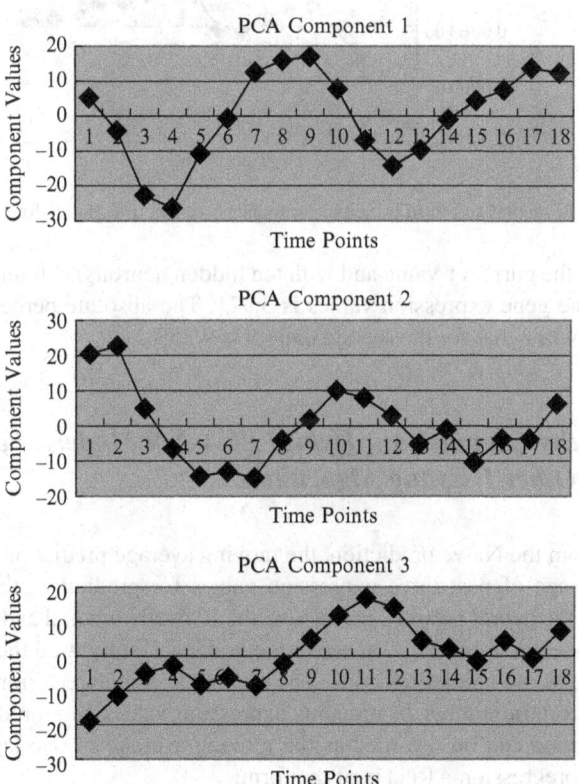

Fig. 5.2 The first, second and third principal components versus time points for Spellman's dataset

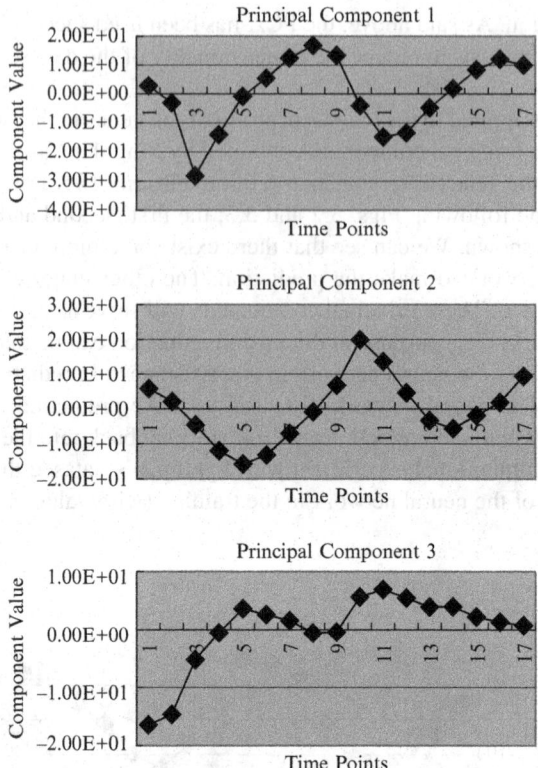

Fig. 5.3 The first, second and third principal components versus time points for Cho's dataset

for predicting the current t-value and with ten hidden neurons) is found to be 1,087. The sum of the gene expression values is 3,921. The absolute percentage error is thus 27.77%, while that for the second dataset is 9.25%.

5.5.3 Results Comparison: Hybrid PCA-NN Models' Performance and Other Existing Algorithms

The results from the Naïve prediction, the moving average prediction (MA), which takes the average of past three expression values for prediction, the autoregression (AR(1)), the neural network prediction, the ICA-NN method and the PCA-NN method are listed in Table 5.1. The naïve method has simply used the previous expression value as the prediction value. The moving average prediction has used the average of a certain number of previous expression values as a predictor. In fact, the Naïve method can be regarded as the moving average of one-lag model. The first-order autoregression AR(1) is of the form:

$$x_t = \rho x_{t-1} + \varepsilon_t$$

Table 5.1 Prediction results from the different methods

Results	Naïve	MA	AR	NN	ICA-NN	PCA-NN
Abs. error (first set)	94.92%	125.87%	80.34%	75%	83.93%	51.31%
Abs. error (second set)	28.52%	39.16%	27.31%	22.52%	27.67%	12.91%

where x_t is the expression level at time t, ρ the coefficient of x_{t-1}. ε_t is the white noise time series with $E[\varepsilon_t] = 0$, $E[\varepsilon_t^2] = \sigma_\varepsilon^2$, and $Cov[\varepsilon_t, \varepsilon_x] = 0$ for all $s \neq t$. These three methods are popular in continuous numerical predictions and their corresponding errors here are the in-sample errors. It can be observed that the NN model and the PCA-NN model are better than these methods. We will see later that the performance of the out-of-sample testing of the PCA-NN model is better than these methods' in-sample testing. The in-sample tests mean that the data is used for both training and testing while the out-of-sample tests mean that the testing data has not been employed in the training process.

5.5.4 Analysis on the Network Structure and the Out-of-Sample Validations

Different combinations of the lag patterns for the training of the neural network with one hidden layer of ten hidden neurons have been tested. Different network architectures have also been checked for comparing their performances.

In Table 5.2, the results of feeding the neural network of ten hidden neurons with different input lag lengths are shown. NN1 represents network with the one-lag model of value t-1, NN2 with the two-lag model of values t-1 and t-2, NN3 with inputs t-1, t-2 and t-3, NN4 with input t-1, t-2, t-3 and t-4, NN5 with input t-1, t-2, t-3, t-4 and t-5.

Table 5.3 shows us the performances of the different network architectures. All are of three-layer structure. NN_10 is with 10 hidden neurons, NN_5 with 5 hidden neurons, and NN_20 with 20 hidden neurons.

The AIC results are also listed in Tables 5.2, 5.3 and 5.4. It has been shown that Akaike's criterion is asymptotically equivalent to the use of cross-validation [39]. Akaike's information criterion (AIC) is defined as:

$$AIC = T \ln(residual\ sum\ of\ squares) + 2n \qquad (5.1)$$

where n is the number of parameters estimated, and T the number of usable observations [19]. The first term of the above AIC equation is to measure the residual sum of squares, and the second term is a penalty for increasing the number of parameters in the model. While a more parsimonious model has the effect of reducing the residual sum of squares, the AIC test can give us a selection criterion that trades off a reduction in the sum of squares of the residuals for a more parsimonious model. We can use the AIC to aid in the selection of the most appropriate model, which is the model with the smallest AIC value (note that it can be negative).

Table 5.2 Prediction results with different input lag lengths for stand-alone neural network

Results	NN1	NN2	NN3	NN4	NN5
Abs. error (first set)	66.23%	63.89%	59.68%	57.82%	55.85%
Abs. error (second set)	19.07%	16.68%	14.76%	13.42%	11.77%
AIC (first set)	189.39	208.31	225.86	245.45	263.63
AIC (second set)	375.13	391.73	409.38	425.41	443.31

Table 5.3 Prediction results with different stand-alone neural network structures

Results	NN_5	NN_10	NN_20
Abs. error (first set)	68.63%	59.68%	47.47%
Abs. error (second set)	18.53%	14.76%	10.93%
AIC (first set)	180.28	225.86	318.77
AIC (second set)	364.50	409.38	500.13

Table 5.4 Prediction results for PCA-NN method with different neural network structures

Results	T1N5	T2N5	T3N5	T1N10	T2N10	T3N10
Abs. error (first set)	67.63%	51.31%	46.85%	57.97%	44.22%	30.99%
Abs. error (second set)	22.02%	12.91%	14.65%	19.01%	8.68%	9.47%
AIC (first set)	160.61	160.87	168.77	184.81	195.75	204.74
AIC (second set)	348.91	338.61	351.59	372.42	367.08	389.33

Table 5.5 Prediction results (abs. percentage error of in-samples and out-of-sample cross-validation) for PCA-NN method with two inputs t-1 and t-2 for neural network structure of five hidden neurons

Results	In-1	Out-1	In-2	Out-2
Spellman's dataset	63.48%	70.86%	66.53%	74.04%
Cho's dataset	19.91%	20.18%	16.02%	24.66%

From the AIC results, the architecture of the t-1 input with five hidden neurons is suggested, as it has the smallest AIC value. And, feeding the neural network of five hidden neurons with input t-1, the error in the first data set is found to be 75% while that in the second data set is 22.52%, which are better than the Naïve, the MA, and the AR methods.

Table 5.4 shows us the prediction results and the AIC values for the PCA-NN method with different network structures. T1 is for feeding the network with input t-1 term only, T2 with input t-1 and t-2 terms, T3 with input t-1, t-1 and t-3 terms. N5 refers to the network of five hidden neurons and N10 of ten neurons. While for the data set 1, the AIC values of TIN5 and T2N5 are more or less the same, the AIC values of the second data set suggest clearly that the model T2N5 should be employed. This AIC result is slightly different from that of the stand-alone neural network.

Table 5.5 lists the results of the cross-validation of the PCA-NN method. The gene expression data is divided into two equal parts. In the first round, In-1 and

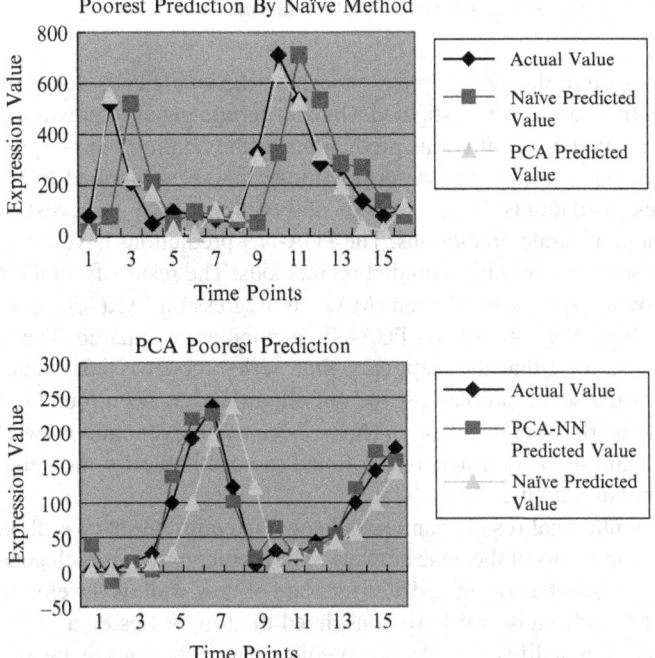

Fig. 5.4 The poorest prediction cases made by Naïve method and PCA-NN method for Cho's dataset (the *top* one is for the gene YDL227c and the *bottom* one for the gene YHL028w)

Out-1 are the prediction errors with the first half as the in-sample data and the second half as out-of-sample data. The In-2 and Out-2 are the results with the opposite partitions. From the table, we can find that the in-sample results are generally better than the out-of-sample results but they are more or less consistent and close to each other. Another observation is that the algorithm's performance of both in-sample and out-of-sample testing is better than other methods' in-sample prediction.

The Fig. 5.4 shows us the poorest prediction results of the Naïve method and the PCA-NN method for Cho's dataset. Among the predictions with the Naïve method, the YDL227c gene expression data has the poorest result, with 65.54% average error. That for the PCA-NN method with the T3N10 model is the gene YHL028W, with 25.45% error. The best prediction result for Naïve method is the gene YDL198c, with 11.62% error. That for PCA-NN prediction result is the gene YBR104w, with only 1.87% error. We can observe that in both cases, the PCA-NN model performs much better that the Naïve method.

Furthermore, the possibility of replacing the PCA with the ICA for the modeling has been tested. It is interesting to note that the results suggest that ICA does not supplement well with neural network for the gene expression time series modeling here.

5.6   Result Discussions and Conclusion

The relationship of the gene expression level in the whole cell cycle from a different prospective has been investigated. Given a certain genes expression profile, the possibility of making continuous predictions of the coming gene expression level changes has been tested. The described model has been applied to two popular gene expression datasets, and it has been observed that PCA can assist the network to make more accurate predictions. The PCA-NN predictions have been compared with other popular continuous prediction methods. The results from the Naïve prediction, moving average prediction (MA), autoregression (AR(1)), neural network prediction, ICA-NN method and PCA-NN method are compared. The autoregression results suggest that the stationary time series model is not suitable for this microarray time series problem, while our algorithm has been found to model the genomic network more accurately. This is because the genomic network is never truly linear and the neural network is found to be suitable for the gene modeling with continuous outputs.

The computational results can let us know the contribution of each gene to the principal components of the gene network. The predictability of each gene's expression value can also be considered as a measure of how well its development can be understood. It is because we have considered the time series data of the gene expression in its whole life cycle. A good prediction model means that we can identify the correct principal component for influencing the gene's developments.

The neural network has been known for its non-linear function capability. Its prediction error is quite reasonable, which is better than the other methods like the Naïve method and the AR method. From the results of the two popular gene expression datasets, we can see that the PCA can assist the neural network to make more accurate predictions and the PCA-NN method outperforms others. A main difficulty in our numeral prediction is that the time points in one cell cycle are short. The changes of the expression levels are very large between each time interval. In short, we need to do further work on this short multivariable time series analysis of the yeast's cell cycle in order to further improve the prediction results. Our system can also be seen as a nonlinear gene inference network. It can give us more accurate model of the genome network, which is never truly linear, while a large-scale gene expression predictive model can obviate the need for an exact understanding of the system at the biochemical level [18].

Genetic algorithm is a promising tool for the optimization of the gene weightings as pointed by Keedwell and Narayanan [30]. Similarly, we can regard our NN numerical prediction as the fitness function of the GA. We are going to select the most influential genes for each gene's development in its life cycle with the GA. The experimental results here have clearly shown that our proposed PCA-NN outperforms the other methods of linear regression, simple neural network and ICA-NN etc. Thus, the suitable candidate to work with the GA will be the PCA-NN model, forming the hybrid GA-PCA-NN system. Another potential method is the ensemble learning, which have been successfully applied for the classification problems in microarray (for example, [48]). Our goal is to achieve a nonlinear gene network that can utilize the microarray data fully, with continuous inputs and continuous outputs, and that can provide us the details of the genes' developmental dependencies. This can be helpful for drug development of the enhancing or inhibiting of a specific gene.

References

1. Aach, J. and Church, G. 2001. Aligning gene expression time series with time warping algorithms. Bioinformatics, 17(6): 495–508.
2. Acta, A. 2001. Chemometric strategies for normalisation of gene expression data obtained from cDNA microarrays. Analytica Chimica Acta, 446(1–2): 449–464.
3. Ao, S. et al. 2006. Gene expression time series modeling with principal component and neural network. Soft Computing – A Fusion of Foundations, Methodologies and Applications, 10(4): 351–359.
4. Ao, S. 2008. Data Mining and Applications in Genomics. Springer: Netherlands.
5. Barron, A. R. 1991. Complexity regularization with application to artificial neural networks. Nonparametric Functional Estimation and Related Topics, 561–576.
6. Barron, A. R. 1992. Neural net approximation. Proceedings of the Seventh Yale Workshop on Adaptive and Learning Systems. New Haven, CT: Yale University Press, pp. 69–72.
7. Bergeron, B. 2003. Bioinformatics Computing. Upper Saddle River, NJ: Prentice Hall.
8. Bicciato et al. 2003. PCA disjoint models for multiclass cancer analysis using gene expression data. Bioinformatics, 19(5): 571–578.
9. Butte, A. et al. 2001. Comparing the similarity of time-series gene expression using signal processing metrics. Journal of Biomedical Informatics, 34: 396–405.
10. Causton, H. et al. 2003. Microarray Gene Expression Data Analysis: A Beginner's Guide. Malden, MA: Blackwell.
11. Chen, T., Filkov, V. and Skiena, S. 2001. Identifying gene regulatory networks from experimental data. Parallel Computing, 27: 141–162.
12. Chen, X. et al. 2008. Supervised principal component analysis for gene set enrichment of microarray data with continuous or survival outcomes. Bioinformatics, 24(21): 2474–2481.
13. Cho, R. et al. 1998. A genome-wide transcriptional analysis of the mitotic cell cycle. Molecular Cell, 2: 65–73.
14. Comon, P. 1994. Independent component analysis-a new concept? Signal Processing, 36: 287–314.
15. Costa, I. G. et al. 2002. A symbolic approach to gene expression time series analysis. Neural Networks 2002 Brazilian Symposium. 25–30.
16. Craig, P., Kennedy, J. and Cumming, A. 2002. Towards visualising temporal features in large scale microarray time-series data. Proceedings of the Sixth International Conference on Information Visualisation – IV2002, London: UK, 10–12 July 2002. 427–433.
17. Dewey, T. 2002. From microarrays to networks: mining expression time series. Information Biotechnology Supplement, Drug Discovery Today, 7(20): 170–175.
18. D'haeseleer, P., Liang, S. and Somogyi, R. 1999. Gene expression data analysis and modeling. Pacific Symposium on Biocomputing, Hawaii: USA.
19. Enders, W. 1995. Applied Econometric Time Series. New York: Wiley.
20. Futschik, M. and Kasabov, N. 2002. Fuzzy clustering of gene expression data. Fuzzy Systems, 2002. Proceedings of the 2002 IEEE International Conference on FUZZ-IEEE'02, Volume: 1, 12–17 May 2002. 414–419.
21. Haykin, S. 1994. Neural Networks: A Comprehensive Foundation. Englewood Cliffs, NJ: Prentice Hall.
22. Haykin, S. 1999. Neural Networks – A Comprehensive Foundation (2nd ed.). Upper Saddle River, NJ: Prentice Hall.

23. Herrero, J., Valencia, A. and Dopzao, J. 2001. A hierarchical unsupervised growing neural network for clustering gene expression patterns. Bioinformatics, 17(2): 126–136.
24. Hornquist, M., Hertz, J. and Wahde, M. 2003. Effective dimensionality of large-scale expression data using principal component analysis. BioSystem, 65: 147–156.
25. Huang, S. H., Tan K. K. and Tang, K. Z. 2004. Neural network control: theory and applications. RSP, Hertfordshire: UK.
26. Hyvärien, A., Karhunen, J. and Oja, E. 2001. Independent Component Analysis. New York: Wiley.
27. Jiang, D., Pei, J. and Zhang, A. 2003. DHC: a density-based hierarchical clustering method for time series gene expression data. Proceedings of the Third IEEE Symposium on Bioinformatics and Bioengineering, Bethesda, Maryland: USA, 10–12 March 2003. 393–400.
28. Ji, X. et al. 2003. Mining gene expression data using a novel approach based on Hidden Markov Models. FEBS Letter, 542: 124–131.
29. Jutten, C. and Herault, J. 1991. Blind separation of sources, part I: and adaptive algorithm based on neuromimetic architecture. Signal Processing, 24: 1–10.
30. Keedwell, E. and Narayanan, A. 2002. Genetic algorithms for gene expression analysis. First European Workshop on Evolutionary Bioinformatics, 2611: 76–86.
31. Kesseli, J., Ramo, P. and Yli-Harja, O. 2004. Inference of Boolean models of genetic networks using monotonic time transformations. First International Symposium on Control, Communications and Signal Processing, Hammamet: Tunisia, 21–24 March 2004. 759–762.
32. Khan, J. et al. 2001. Classification and diagnostic prediction of cancers using gene expression profiling and artificial neural networks. Nature Medicine, 7(6): 673–679.
33. Kong, W. et al. 2009. Independent component analysis of Alzheimer's DNA microarray gene expression data. Molecular Neurodegener, 4: 5, doi: 10.1186/1750-1326-4-5.
34. Langmead, C., McClung, C. and Donald, B. 2002. A maximum entropy algorithm for rhythmic analysis of genome-wide expression patterns. Bioinformatics Conference 2002, IEEE. 237–245.
35. Lynn, K. et al. 2009. A neural network model for constructing endophenotypes of common complex diseases: an application to male young-onset hypertension microarray data. Bioinformatics, 25(8): 981–988.
36. McCulloch, W. W. and Pitts, W. 1943. A logical calculus of the ideas imminent in nervous activity. Bulletin of Mathematical Biophysics, 5: 115–133.
37. Nikkilä, J. et al. 2002. Analysis and visualization of gene expression data using self-organizing maps. Neural Networks, 15(8–9): 953–966.
38. Peterson, C. and Ringner, M. 2003. Analyzing tumor gene expression profiles. Artificial Intelligence in Medicine, 28: 59–74.
39. Principe, J., Euliano, N. and Lefebvre, W. 2000. Neural and Adaptive Systems: Fundamentals Through Simulations. New York: Wiley.
40. Rosenblatt, F. 1958. The perceptron: a probabilistic model for information storage and organization in the brain. Psychological Review, 65(6): 386–408.
41. Sakamoto, E. and Iba, H. 2001. Inferring a system of differential equations for a gene regulatory network by using genetic programming. Proceedings of the 2001 Congress on Evolutionary Computation, Volume: 1, 27–30 May 2001. 720–726.
42. Sawa, T. and Ohno-Machado, L. 2003. A neural network-based similarity index for clustering DNA microarray data. Computers in Biology and Medicine, 33: 1–15.
43. Spellman, P. et al. 1998. Comprehensive identification of cell cycle-regulated genes of the yeast saccharomyces cerevisiae by microarray hybridization. Molecular Biology of the Cell, 9: 3273–3297.
44. Spicker, J. et al. 2002. Neural network predicts sequence of TP53 gene based on DNA chip. Bioinformatics, 18(8): 1133–1134.
45. Syeda-Mahmood, T. 2003. Clustering time-varying gene expression profiles using scale-space signals. Proceedings of the 2003 IEEE Computer Society Bioinformatics Conference, Stanford, CA: USA, 11–14 Aug. 2003. 48–56.

46. Tabus, I. and Astola, J. 2003. Clustering the non-uniformly sampled time series of gene expression data. Proceedings of the Seventh International Symposium on Signal Processing and Its Applications, Volume: 2, 1–4 July 2003. 61–64.
47. Tabus, I., Giurcaneanu, C. and Astola, J. 2004. Genetic networks inferred from time series of gene expression data. First International Symposium on Control, Communications and Signal Processing, Hammamet: Tunisia, 21–24 March 2004. 755–758.
48. Tan, A. and Gilbert, D. 2003. Ensemble machine learning on gene expression data for cancer classification. Applied Bioinformatics, 2(3 Suppl.): S75–S83.
49. Taylor, J. et al. 2002. Application of metabolomics to plant genotype discrimination using statistics and machine learning. Bioinformatics, 18(Suppl. 2): 241–248.
50. Theodoridis, S. and Koutroumbas, K. 2003. Pattern Recognition (2nd ed.). San Diego, CA: Academic Press.
51. Veiga, D. et al. 2008. Predicting transcriptional regulatory interactions with artificial neural networks applied to E. coli multidrug resistance efflux pumps. BMC Microbiology, 8:101. doi: 10.1186/1471-2180-8-101.
52. Vogl, T. et al. 1998. Accelerating the convergence of the backpropagation method. Biological Cybernetics, 59: 257–263.
53. Widrow, B. 1959. Generalization and information storage in networks of adaline neurons. In Yovits, M.C., Jacobi, G.T. and Goldstein, G.D. (Eds.), Self-Organizing Systems. Washington, DC: Spartan, 435–461.
54. Wolkenhauer, O. 2002. Mathematical modeling in the post-genome era: understanding genome expression and regulation-a system theoretic approach. BioSystems, 65: 1–18.
55. Wu, F., Zhang, W. and Kusalik, A. 2003. Determination of the minimum sample size in microarray experiments to cluster genes using k-means clustering. Proceedings of the Third IEEE Symposium on Bioinformatics and Bioengineering, Bethesda, Maryland: USA, 10–12 March 2003. 401–406.
56. Yeang, C. and Jaakkola, T. 2003. Time series analysis of gene expression and location data. Proceedings of the Third IEEE Symposium on Bioinformatics and Bioengineering, Bethesda, Maryland: USA, 10–12 March 2003. 305–312.
57. Yeung, K. and Ruzzo, W. 2001. Principal component analysis for clustering gene expression data. Bioinformatics, 17(9): 763–774.
58. Yoshioka, T. and Ishii, S. 2002. Clustering for time-series gene expression data using mixture of constrained PCAS. Neural Information Processing, ICONIP '02. 2239–2243 (v5).
59. Yukalov, V. 2000. Self-similar extrapolation of asymptotic series and forecasting for time series. Modern Physics Letters B, 14 (22/23): 791–900.
60. Zhang, L., Zhang, A. and Ramanathan, M. 2003. Fourier harmonic approach for visualizing temporal patterns of gene expression data. Proceedings of the 2003 IEEE Computer Society Bioinformatics Conference, Stanford, CA: USA, 11–14 Aug. 2003. 137–147.

Chapter 6
An Asymptotic Method to a Financial Optimization Problem

Dejun Xie, David Edwards, and Giberto Schleiniger

Abstract This paper studies the borrower's optimal strategy to close the mortgage when the volatility of the market investment return is small. Integral equation representation of the mortgage contract value is derived, then used to find the numerical solution of the free boundary. The asymptotic expansions of the free boundary are derived for both small time and large time. Based on these asymptotic expansions two simple analytical approximation formulas are proposed. Numerical experiments show that the approximation formulas are accurate enough from practitioner's point of view.

Keywords Mortgage prepayment · Asymptotic analysis · Numerical solution · Analytical approximation

6.1 Introduction

Consider a mortgage with a fixed interest rate of c (year^{-1}). Assume that the underlying risk free rate following the CIR model [1], which says $dr_t = k(\theta - r_t)dt + \sigma\sqrt{r_t}dW_t$, where k, θ, σ are positive constants. According to standard mathematical finance theory (see [4, 8–10], for instance), the value of the mortgage contract $V(x,t)$ at any specified t, the time left to the expiry of the contract, and the corresponding interest rate x, when it is not optimal for prepayment, satisfies

$$\frac{\partial V}{\partial t} - \frac{\sigma^2}{2}x\frac{\partial^2 V}{\partial x^2} - k(\theta - x)\frac{\partial V}{\partial x} + xV = m; \qquad (6.1)$$

and when the borrower decides to terminate the contract prematurely at time t, he needs to pay the mortgage loan balance

$$M(t) = \frac{m}{c}\left[1 - e^{-ct}\right], \qquad (6.2)$$

D. Xie (✉), D. Edwards, and G. Schleiniger
Department of Mathematics, University of Delaware
e-mail: dxie@UDel.Edu

S.-I. Ao et al. (eds.), *Advances in Machine Learning and Data Analysis*,
Lecture Notes in Electrical Engineering 48, DOI 10.1007/978-90-481-3177-8_6,
© Springer Science+Business Media B.V. 2010

where m denotes the continuous mortgage payment rate, i.e., the borrower pays mdt (dollars) to the mortgage contract holder (the lender) for each time period dt. Mathematically we have a free boundary problem where the free boundary $x = h(t)$ defines the optimal market interest rate level at which the borrower should terminate the contract. For the continuation region where $x > h(t)$, the contract is in effect and the value of the contract satifies (6.1). For the early exercise region where $x \leq h(t)$, the contract is closed and the lender gets back the loan balance of $M(t)$. Because it is the borrower, rather than the lender, who is a proactive player of the game and has the choice to act in response to the market, so the value of the contract is always less or equal than the loan balance. Thus the free boundary is where the value of the contract $V(x, t)$ first reaches the value of the mortgage loan balance $M(t)$. It is easy to show, using the free of arbitrage argument, that the free boundary starts from c, i.e., $h(0) = c$. And because of the smooth patch is needed for the regularity of the problem, we have the derivative of $V(x, t)$ must be 0 on $h(t)$. Lastly, it is trivially true that $V(x, 0) = 0$, which says that the value of the contract, when the contract is expired, must be 0. Putting all these condition together, we formulate the problem as follows: for $\forall x \geq 0$ and $t > 0$, find $V(x, t)$ and $h(t)$ such that

$$
\begin{cases}
\mathbf{L}(V) = m, & \text{for } x > h(t), t > 0 \\[2mm]
V = \dfrac{m}{c}[1 - e^{-ct}], & \text{for } x \leq h(t), t > 0 \\[2mm]
\dfrac{\partial V}{\partial x}(h(t), t) \equiv 0 \\[2mm]
V(x, 0) = 0, & \text{for all } x \geq 0 \\[2mm]
h(0) = c
\end{cases}
\tag{6.3}
$$

where the differential operator $\mathbf{L}$ is defined as

$$
\mathbf{L}(V) = \frac{\partial V}{\partial t} - \frac{\sigma^2}{2} x \frac{\partial^2 V}{\partial x^2} - k(\theta - x) \frac{\partial V}{\partial x} + x V
\tag{6.4}
$$

Because of the important role played by mortgage backed securities in real economy, there has been continuing interest in mortgage pricing and related problems, especially the prepayment strategies for mortgage borrowers. Most of the studies, such as [2, 5], are from option-theoretical viewpoint. A similar problem with underlying interest rate following Vasicek model was recently studied with variational integral equation approach in [3, 7]. In this paper, we focus on the situation where the volatility σ is small. Such an assumption is reasonable for the long term real economy. More discussions on parameter estimation for risk free market return can be found in [6].

Here we first derive the integral representation of the solution with the free boundary embedded, then prove the monotonocity and boundedness of the free boundary, then design an effective iteration scheme to solve the problem numerically. Based on the asymptotic analysis, we drive two analytical approximation formulas for the optimal prepayment boundary. Numerical simulations are carried out to validate our approximation formulas.

6.2   Integral Representation of the Solution

We first derive the characteristic solution. When $\sigma \to 0$, the PDE (6.1) reduces to

$$\frac{\partial V}{\partial t} - k(\theta - x)\frac{\partial V}{\partial x} + xV = m. \tag{6.5}$$

Lemma 6.2.1 *The characteristic solution associated with (6.5) is*

$$V(X(t),t) = me^{-\theta t - \frac{X(t)-\theta}{k}} \int_0^t e^{\theta \tau + \frac{X(\tau)-\theta}{k}} d\tau, \tag{6.6}$$

where

$$X(t) = \theta + (X_0 - \theta)e^{kt} \tag{6.7}$$

for each given $X(0) = X_0$.

Proof. Starting with each intial point $(X(0),0)$, $X(0) = X_0 \geq 0$, we define the characteristic curve related to (6.5) as

$$\frac{\partial X}{\partial t} = -k(\theta - X), \quad X(0) = X_0,$$

which gives

$$X(t) = \theta + (X_0 - \theta)e^{kt}. \tag{6.8}$$

Now on each characteristic curve, we have the following ODE for $V(X(t),t)$ defined as

$$\frac{dV}{dt} + (\theta + (X_0 - \theta)e^{kt})V = m$$

or equivalently

$$\frac{d}{dt}\left\{Ve^{\theta t + \frac{X_0-\theta}{k}e^{kt}}\right\} = me^{\theta t + \frac{X_0-\theta}{k}e^{kt}}$$

the solution of which is

$$V = me^{-\theta t - \frac{X_0-\theta}{k}e^{kt}} \int_{t^*}^t e^{\theta \tau + \frac{X_0-\theta}{k}e^{k\tau}} d\tau \tag{6.9}$$

Because of the requirement of $V(X(0),0) = 0$, we have $t^* = 0$.

Lemma 6.2.2 *The solution to (6.5) is given by*

$$V(x,t) = me^{-\frac{x-\theta}{k}} \int_0^t e^{-\theta s + \frac{x-\theta}{k}e^{-ks}} ds, \tag{6.10}$$

it is strictly decreasing in x, ranges from $\lim_{x \to -\infty} V(x,t) = \infty$ to $\lim_{x \to \infty} V(x,t) = 0$.

Proof. Reformulate the equation of $V(X(t),t)$ in (6.6), we have

$$V(X(t),t) = me^{-\theta t - \frac{X(t)-\theta}{k}} \int_0^t e^{\theta \tau + \frac{X(\tau)-\theta}{k}} d\tau$$

$$= me^{-\frac{X(t)-\theta}{k}} \int_0^t e^{-\theta s + \frac{X(t)-\theta}{k} e^{-ks}} ds.$$

Write $x = X(t)$, we have the desired form of the solution as in (6.10). To validate the monotonocity of V in x, we note that, in the set where V satisfies (6.5), $V_x := \frac{\partial V}{\partial x}$ satisfies the differential inequality

$$\frac{\partial V_x}{\partial t} - k(\theta - x)\frac{\partial V_x}{\partial x} + (x+k)V_x = -V < 0.$$

The limits of V as x approaches $\pm\infty$ are the results of a simple computation.

6.3 Properties of the Free Boundary

In this section we shall show the monotonocity of the free boundary $h(t)$ and the existence of $\lim_{t \to \infty} h(t)$ and $\lim_{t \to 0_+} h(t)$, namely, we shall prove the following theorem:

Theorem 6.1. *If* $c < \theta$, *then* $h(t)$, *starts from* $h(0) = c$, *is continuous and monotonously decreasing in* $[0, \infty)$, *and is lower bounded. If* $c > \theta$, *then* $h(t)$, *starts from* $h(0) = c$, *is continuous and monotonously increasing in* $[0, \infty)$, *and is upper bounded.*

Proof. The theorem is a summary of the following Lemmas 6.3.1–6.3.6 and Corollaries 1–2. The proof is organized as follows: we first show the existence, uniqueness, and continuity of $h(t)$, except possibly for $t = 0$, then show the boundedness of $h(t)$ both from below and above, then the monotonocity, and lastly we find the limit of $h(t)$ at $t = 0$.

Lemma 6.3.1 *For each* $t \geq 0$, $h(t)$ *exists and is unique.* $h(t)$ *is continuous for all* $t \geq 0$ *except possibly at* $x = 0$.

Proof. The existence and uniqueness is naturally concluded from Lemma 6.2.2. The continuity of $h(t)$ for $t > 0$ is a consequence of the continuity of V in x. The only thing left to validate is $\lim_{t \to 0_+} h(t) = c$, which is to be done after we prove the boundedness of $h(t)$.

Lemma 6.3.2 *If* $c > \theta$, $\sigma \to 0$, *the free boundary* $h(t)$ *in (6.3) is lower bounded by* c, *i.e.,*

$$h(t) > c \quad \forall t > 0. \tag{6.11}$$

Proof. Because $V(X(t), t)$ is monotoneously decreasing (to 0) in $X(t)$ for fixed $t > 0$, i.e., $\frac{\partial V}{\partial X} < 0, \forall t > 0$, which is shown in the Lemma 6.2.2, it suffices to show $V(c, t) > M(t)$, where $M(t) = \frac{m}{c}\left[1 - e^{-ct}\right]$ is the contract value on the free boundary. Recall $V(x, t) = e^{-\frac{x-\theta}{k}} \int_0^t e^{-\theta s + \frac{x-\theta}{k} e^{-ks}} ds$ (hereafter, we assume, WLOG, $m = 1$), we have

$$V(c, t) = e^{-\alpha} \int_0^t e^{(k\alpha - c)s + \alpha e^{-ks}} ds,$$

by letting $\frac{c-\theta}{k} = \alpha$. Now, noticing $\alpha > 0$, we have

$$V(c, t) - M(t) = e^{-\alpha}\left\{\int_0^t e^{-cs}\left[e^{k\alpha s + \alpha e^{-ks}} - e^{\alpha}\right] ds\right\}.$$

Because

$$k\alpha s + \alpha e^{-ks} = \alpha(ks + e^{-ks})$$
$$> \alpha$$

We have

$$V(c, t) - M(t) > 0$$

and thus completes the proof.

Corollary 6.1. *If $c < \theta$, $\sigma \to 0$, the free boundary $h(t)$ in (6.3) is upper bounded by c, i.e.,*

$$h(t) < c \quad \forall t > 0.$$

Proof. Follow the same procedure of the above proof except this time $\alpha < 0$, and thus changes the sign of $V(c, t) - M(t)$.

Lemma 6.3.3 *If $c > \theta$, then $h(t)$ is monotonously increasing in t, i.e., $h'(t) > 0, \forall t > 0$, and $\lim_{t \to \infty} h'(t) = 0$.*

Proof. Knowing that $V(h(t), t) = \frac{1}{c}\left[1 - e^{-ct}\right]$, we have, for $\forall t > 0$,

$$e^{-\frac{h(t)-\theta}{k}} \int_0^t e^{-\theta s + \frac{h(t)-\theta}{k} e^{-ks}} ds = \frac{1}{c}\left[1 - e^{-ct}\right],$$

or equivalently,

$$\int_0^t e^{-\theta s - \frac{h(t)-\theta}{k}[1-e^{-ks}]} ds = \frac{1}{c}\left[1 - e^{-ct}\right].$$

Differentiating it with respect to t, we get

$$-\frac{h'(t)}{k}\int_0^t e^{-\theta s - \frac{h(t)-\theta}{k}[1-e^{-ks}]}\left[1 - e^{-ks}\right] ds$$

$$+ e^{-\theta t - \frac{h(t)-\theta}{k}[1-e^{-kt}]} = e^{-ct}. \tag{6.12}$$

Notice that the definite integral in above equation is strictly positive. If the second term is strictly greater than e^{-ct}, then $h'(t) > 0$ is necessary for the above equation to hold. Now the previous Lemma 6.3.2 tells us that $h(t) > c$, hence

$$h(t) - \theta > 0,$$

and also

$$0 < 1 - e^{-kt} < kt,$$

we have

$$\frac{h(t) - \theta}{k}[1 - e^{-kt}] < \frac{c - \theta}{k}kt = ct - \theta t.$$

So

$$e^{-\theta t - \frac{h(t)-\theta}{k}[1-e^{-kt}]} > e^{-\theta t - (ct - \theta t)} = e^{-ct},$$

which is the desired inequality leading the monotonocity of the $h(t)$. Lastly, if we let $t \to \infty$ in (6.3), we have both the righthand side and the second term in the left side vanish, thus forces the first term in the left side vanish too. But the definite integral itself is strictly positive, so $\lim_{t\to\infty} h'(t) = 0$ becomes necessary, thus completes the proof.

Corollary 6.2. *If $c < \theta$, then $h(t)$ is monotonously decreasing in t, i.e., $h'(t) < 0, \forall t > 0$, and $\lim_{t\to\infty} h'(t) = 0$.*

Lemma 6.3.4 *If $c > \theta$, then $\lim_{t\to\infty} h(t)$ exists. For $\forall$ fixed $\epsilon > 0$, $\lim_{t\to\infty} h(t) < \left[c - \theta + \frac{c}{\theta}e^{-(\epsilon\theta+1)}\right]\frac{\epsilon k}{1 - e^{-\epsilon k}} + \theta$. In particular, $\lim_{t\to\infty} h(t) < c + \frac{c}{\theta}$*

Proof. Let $\lim_{t\to\infty} h(t) = h^*$. Knowing the contract value at t infinity is $\frac{1}{c}$, we wish to balance the following parametric integral of h^*

$$\frac{1}{c} = \int_0^\infty e^{-\theta s - \frac{h^*-\theta}{k}[1-e^{-ks}]} ds.$$

The boundedness of h^* is immediate simply because $\lim V(x,t)_{x\to\infty} \to 0$. Here we are interested in finding a particular value of the bound. Fix $\epsilon > 0$, let $1 - e^{-k\epsilon} = \lambda$. Notice that $1 - e^{-ks} > \frac{\lambda}{\epsilon}$ for $0 < s < \epsilon$ and $1 - e^{-ks} > \lambda$ for $s > \epsilon$, we have

$$\frac{1}{c} < \int_0^\epsilon e^{-\theta s - \frac{h^*-\theta}{k}\frac{\lambda}{\epsilon} s} ds + \int_\epsilon^\infty e^{-\theta s - \frac{h^*-\theta}{k}\lambda} ds = \frac{\theta + ye^{-(\theta+y)\epsilon}}{(\theta + y)\theta},$$

where $y := \frac{h^*-\theta}{k}\frac{\lambda}{\epsilon}$. Now we have

$$\frac{\theta(c-\theta)}{c} > \frac{\theta}{c}y - ye^{-(\theta+y)\epsilon},$$

since $c > \theta$. The condition $h^* > c > \theta$ here plays its role because otherwise $\theta + y$ is not necessarily positive. Notice that the function defined by $f(y) = ye^{-(\theta+y)\epsilon}$ achieves the absolute maximum of $e^{-(\epsilon\theta+1)}$ at $y = \frac{1}{\epsilon}$, we have

$$\frac{\theta(c-\theta)}{c} > \frac{\theta}{c}y - e^{-(\epsilon\theta+1)}.$$

Correspondingly, we have

$$h^* < \left[c - \theta + \frac{c}{\theta}e^{-(\epsilon\theta+1)}\right]\frac{\epsilon k}{1 - e^{-\epsilon k}} + \theta.$$

The righthand side of above inequality is continuous in ϵ. Take limit for $\epsilon \to 0$, we find $[c - \theta + \frac{c}{\theta}e^{-(\epsilon\theta+1)}]\frac{\epsilon k}{1-e^{-\epsilon k}} + \theta < c + \frac{c}{\theta}$.

Lemma 6.3.5 *If $c < \theta$, then $\lim_{t\to\infty}h(t)$ exists. For $\forall$ fixed $\epsilon > 0$, $\lim_{t\to\infty}h(t) >$ $\theta\left(1 - \frac{k\epsilon}{1-e^{-k\epsilon}}\right) - \frac{k\theta}{c}\frac{1}{1-e^{-k\epsilon}}$.*

Proof. Again, we wish to balance the following parametric integral of h^*

$$\frac{1}{c} = \int_0^\infty e^{-\theta s - \frac{h^*-\theta}{k}[1-e^{-ks}]} ds.$$

Due to Lemma 6.2.2, the lower boundedness of h^* is apparent. Here we are interested in finding a particular value of the lower bound. Fix $\epsilon > 0$, let $1 - e^{-k\epsilon} = \lambda$. Notice that $1 - e^{-ks} > \frac{\lambda}{\epsilon}$ for $0 < s < \epsilon$ and $1 - e^{-ks} > \lambda$ for $s > \epsilon$, we have

$$\frac{1}{c} > \int_0^\epsilon e^{-\theta s - \frac{h^*-\theta}{k}\frac{\lambda}{\epsilon} s} ds + \int_\epsilon^\infty e^{-\theta s - \frac{h^*-\theta}{k}\lambda} ds$$

$$= \frac{\theta - ye^{-(\theta-y)\epsilon}}{(\theta - y)\theta},$$

where $y := \frac{\theta-h^*}{k}\frac{\lambda}{\epsilon}$. If $\theta - y \geq 0$, then

$$\theta \geq (\theta - h^*) \frac{1 - e^{-k\epsilon}}{k\epsilon},$$

which is equivalent to

$$h^* \geq \theta \left(1 - \frac{k\epsilon}{1 - e^{-k\epsilon}}\right).$$

If $\theta - y < 0$ then we have

$$\frac{1}{c} > \frac{\theta - ye^{-(\theta - y)\epsilon}}{(\theta - y)\theta}$$

$$= \frac{e^{(y-\theta)\epsilon} - 1}{\theta - y} + \frac{e^{(y-\theta)\epsilon}}{\theta}$$

$$> \frac{\epsilon}{\theta}(y - \theta),$$

which gives

$$h^* \geq \theta \left(1 - \frac{k\epsilon}{1 - e^{-k\epsilon}}\right) - \frac{k\epsilon}{1 - e^{-k\epsilon}} \frac{\theta}{\epsilon c}.$$

Because we are looking for a lower bound, so we take the minimum of the two cases (for fixed $\epsilon > 0$), and conclude that

$$h^* \geq \theta \left(1 - \frac{k\epsilon}{1 - e^{-k\epsilon}}\right) - \frac{k\theta}{c} \frac{1}{1 - e^{-k\epsilon}}.$$

Lemma 6.3.6 $h(t)$ *is continuous for* $t \in [0, \infty)$, *in particular,* $\lim_{t \to 0_+} = c$.

Proof. Because of Lemma 6.3.1, the only thing left to be justified is $\lim_{t \to 0_+} = c$. For t small, $e^{-ks} = 1 - ks$, we have

$$\lim_{t \to 0_+} V(h(t), t) = \lim_{t \to 0_+} e^{-\frac{h(t) - \theta}{k}} \int_0^t e^{-\theta s + \frac{h(t) - \theta}{k} e^{-ks}} ds$$

$$= \lim_{t \to 0_+} e^{-\frac{h(t) - \theta}{k}} \int_0^t e^{-\theta s + \frac{h(t) - \theta}{k}(1 - ks)}$$

Because of the continuity and boundedness of $h(t)$, we can take limit of $\lim_{t \to 0_+} h(t)$ inside of the integral and arrive at

$$\lim_{t \to 0_+} V(h(t), t) = \frac{1}{\lim_{t \to 0_+} h(t)}$$

Compare this with the boundary value of $\frac{1}{c}[1 - e^{ct}]$, we have that $\lim_{t \to 0_+} h(t) = c$.

6.4   Numerical Solution of the Free Boundary

Since $\frac{\partial V}{\partial x} \neq 0$ we can use Newton method to solve for the free boundary iteratively. Define

$$Q(h) = e^{-\frac{h-\theta}{k}} \int_0^t e^{-\theta s + \frac{h-\theta}{k} e^{-ks}} ds - \frac{1}{c}\left[1 - e^{ct}\right],$$

and

$$f(h) = e^{-\frac{h-\theta}{k}} \left[-\frac{1}{k}\right] \int_0^t e^{-\theta s + \frac{h-\theta}{k} e^{-ks}} ds$$

$$+ e^{-\frac{h-\theta}{k}} \int_0^t e^{-\theta s + \frac{h-\theta}{k} e^{-ks}} \left[\frac{1}{k} e^{-ks}\right] ds,$$

our problem is to find h such that

$$Q[h](t) \equiv 0, \qquad \forall t \geq 0.$$

For fixed $t = T$, discretize $[0, T]$ uniformly into n subintervals by $t_0, t_1, t_2, ..., t_n$, where $t_0 = 0, t_n = T$. Start with $h(t_0) = c$ and assume $h(t_1), h(t_2), ..., h(t_{n-1})$ are known, to compute $h(t_n)$ with Newton's algorithm, we first assign a reasonable initial guess for $h(t_n)$ as

$$h^0(t_n) = h(t_{n-1}), \qquad n = 1;$$
$$h^0(t_n) = 2h(t_{n-1}) - h(t_{n-2}), \qquad n > 1.$$

For a given error tolerance level, say $Tole = 10^{-7}$, we have the following Newton's iteration scheme

$$h(t_n)^{new} = h(t_n)^{old} - \frac{Q(h(t_n)^{old})}{f(h(t_n)^{old})}.$$

After each step of iteration, a current error is recorded as

$$error(k) = h(t_n)^{new} - h(t_n)^{old}.$$

The iteration is kept running until an integer k is reached such that $error(k) < Tole$. To increase the accuracy of the numerical solution, one can increase N, the number of grids for partitioning the time interval $[0, T]$. For typical parameters with $T \leq 25$, our numerical simulations show that $N = 4096$ is large enough for achieving a solution with relative error less than 10^{-7}, where relative error is defined as the difference of numerical values of $h(T)$'s achieved with different N's. Figure 6.1 is a numerical plot of the free boundaries as we fix one set of parameters at a time

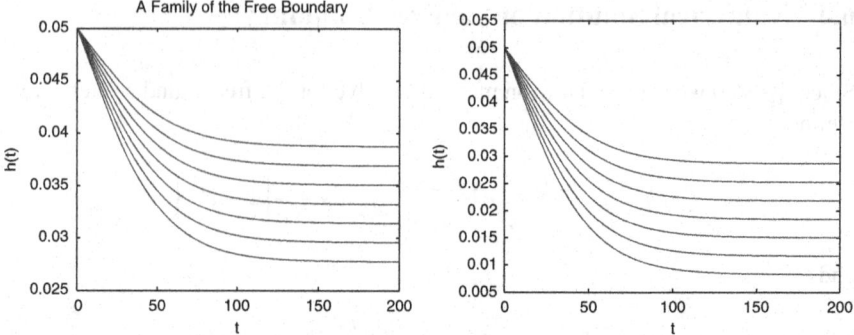

Fig. 6.1 c = 0.05, $k = 0.06, 0.07, ..., 0.12$ (*top to bottom*) $\theta = 0.06$ (*right*), 0.07 (*left*). The units for t and $h(t)$ are years and year^{-1}, respectively

6.5 Asymptotic Analysis of the Free Boundary

In this section, we derive asymptotic expansions of $h(t)$ for both small t and large t.

Theorem 6.2. *As $t \to 0$, $h(t) \sim c + \alpha t$, where $\alpha = \frac{(c-\theta)k}{3}$.*

Proof. We postulate that as $t \to 0$, $h(t) \sim c + \alpha t$, plug this into the contract value on $h(t)$, we have that, for t small,

$$V(h(t), t) = \int_0^t e^{-\theta s - \frac{c+\alpha t - \theta}{k}[1 - e^{-ks}]} ds$$

$$= \int_0^t e^{\frac{(c-\theta+\alpha t)k}{2} s^2 - (c+\alpha t)s} ds$$

For $a, b > 0$, s small, we have the following Taylor expansion

$$e^{as^2 - bs} = \left\{ 1 + as^2 + \frac{a^2 s^4}{2!} + ... \right\} \left\{ 1 - bs + \frac{b^2 s^2}{2!} - ... \right\}$$

$$= 1 - bs + \left(a + \frac{b^2}{2} \right) s^2 - \left(ab + \frac{b^3}{3!} \right) s^3 + o(s^3)$$

Integrating it term by term for small t, we have

$$\int_0^t e^{as^2 - bs} ds = t - \frac{b}{2}t^2 + \frac{1}{3} \left(a + \frac{b^2}{2} \right) t^3$$

$$- \frac{1}{4} \left(ab + \frac{b^3}{3!} \right) t^4 + o(t^4).$$

In terms of our problem at hand, we have

$$V(h(t),t) = t - \frac{c+\alpha t}{2}t^2 + \frac{1}{3}\left[\frac{c-\theta+\alpha t}{2} + \frac{(c+\alpha t)^2}{2}\right]t^3 - o(t^3).$$

We want to match this, term by term, with the known expression of $V(h(t),t)$, which is

$$\frac{1}{c}[1 - e^{-ct}] = t - \frac{c}{2}t^2 + \frac{1}{3!}c^2 t^3 - o(t^3).$$

In the above two series, the coefficients for t and t^2 match each other automatically. To match the t^3 terms, we need to have

$$-\frac{\alpha}{2} + \frac{1}{3}\left(\frac{(c-\theta)k + c^2}{2}\right) = \frac{1}{6}c^2,$$

which gives

$$\alpha = \frac{(c-\theta)k}{3}.$$

Theorem 6.3. *There exist constants* $h^* = \lim_{t\to\infty} h(t)$, $\rho_1 > 0$, *and* $\rho_2 > 0$ *such that, as* $t \to \infty$,

$$h(t) \sim h^* - \rho_1 e^{-\theta t}, \quad if \quad c < \theta,$$

$$h(t) \sim h^* + \rho_2 e^{-ct}, \quad if \quad c > \theta,$$

where h^* *is implicitly given by* $M\left(1, \frac{\theta}{k} + 1, -\frac{h^*-\theta}{k}\right) = \frac{\theta}{c}$, *where* $M(p,q,z)$ *is the confluent hypergeometric function of the first kind of order* p, q, *and*

$$\rho_1 = \frac{kc(h^* - \theta)e^{-\frac{h^*-\theta}{k}}}{\theta(h^* - c)},$$

$$\rho_2 = \frac{k(h^* - \theta)}{h^* - c}.$$

The existence and boundedness of h^* have been previously shown. The main idea to find the exact value of h^* is to use repeated integration by parts to express the contract value V at t infinity as a infinite series involving h^*, which turns out to be a confluent hypergeometric function. As in general, given $a, b, c > 0$, we have

$$\int_0^\infty e^{-ay+be^{-cy}} dy = -\frac{1}{a}\int_0^\infty e^{be^{-cy}}[e^{-ay}]' dy$$

$$= \frac{1}{a}e^b + (-1)\frac{bc}{a}\int_0^\infty e^{-(a+c)y+be^{-cy}} dy$$

Repeating the integration by parts, using the recursive identity

$$\int_0^\infty e^{-(a+nc)y+be^{-cy}}\,dy = -\frac{1}{a+nc}e^b + (-1)\frac{bc}{a+nc}\int_0^\infty e^{-(a+(n+1)y)+be^{-cy}}\,dy,$$

where the tail definite integral vanishes as $n \to \infty$, we have

$$\int_0^\infty e^{-ay+be^{-cy}}\,dy = \frac{1}{a}\sum_{n=1}^{n=\infty}(-1)^n\frac{b^n}{(a/c+1)(a/c+2)...(a/c+n)}e^b$$

$$= \frac{1}{a}M(1, a/c+1, -b).$$

In terms of our problems, this means

$$e^{-\frac{h^*-\theta}{k}e^{-ks}}\int_0^\infty e^{-\theta s+\frac{h^*-\theta}{k}e^{-ks}}\,ds = \frac{1}{\theta}M\left(1,\theta/k+1,-\frac{h^*-\theta}{k}\right).$$

At t infinity, we want

$$V(h(\infty),\infty) = e^{-\frac{h^*-\theta}{k}e^{-ks}}\int_0^\infty e^{-\theta s+\frac{h^*-\theta}{k}e^{-ks}}\,ds = \frac{1}{c},$$

which means

$$M\left(1,\frac{\theta}{k}+1,-\frac{h^*-\theta}{k}\right) = \frac{\theta}{c}.$$

To fully understand the asymptotic behavior of the free boundary as $t \to \infty$, we evaluate the limit of $h'(t)$ as $t \to \infty$. Start with the equation

$$\int_0^\infty e^{-\theta s-\frac{h(t)-\theta}{k}[1-e^{-ks}]}\,ds = \frac{1}{c}\left[1-e^{-ct}\right],$$

take derivative with respect to t along $h(t)$,

$$-\frac{h'(t)}{k}\int_0^t e^{-\theta s-\frac{h(t)-\theta}{k}[1-e^{-ks}]}\left[1-e^{-ks}\right]ds =$$

$$-e^{-\theta t-\frac{h(t)-\theta}{k}[1-e^{-kt}]}+e^{-ct},$$

we get

$$\frac{h'(t)}{k} = \frac{e^{-\theta t-\frac{h(t)-\theta}{k}[1-e^{-kt}]}-e^{-ct}}{\frac{1}{c}[1-e^{-ct}]-I},$$

where

$$I := \int_0^t e^{-\theta s - \frac{h(t)-\theta}{k}[1-e^{-ks}]} e^{-ks} ds,$$

which can be evaluated using integration by parts. Thus we get

$$\frac{h'(t)}{k} = \frac{(h(t)-\theta)\left\{e^{-\theta t - \frac{h(t)-\theta}{k}[1-e^{-kt}]} - e^{-ct}\right\}}{\frac{h(t)}{c}[1-e^{-ct}] + e^{-\theta t - \frac{h(t)-\theta}{k}[1-e^{-kt}]} - 1}.$$

When $c > \theta$, we can write the above equation into

$$h'(t) = F(t)e^{-\theta t},$$

where

$$F(t) = \frac{e^{-\frac{h(t)-\theta}{k}[1-e^{-kt}]} - e^{-(c-\theta)t}}{\frac{h(t)}{c(h(t)-\theta)}[1-e^{-ct}] + \frac{1}{h(t)-\theta}\left\{e^{-\theta t - \frac{h(t)-\theta}{k}[1-e^{-kt}]}\right\}}.$$

Since it has been shown that $\lim_{t\to\infty} h(t) = h^* > \theta$, it is straightforward to show that $F(t)$ is uniformly bounded in t and

$$\lim_{t\to\infty} F(t) = \frac{kc(h^*-\theta)e^{-\frac{h^*-\theta}{k}}}{h^*-c}.$$

Now we postulate

$$h(t) \sim h^* - \rho_1 e^{-\theta t}, \quad \text{if } c > \theta.$$

Compare the limit of $h'(t)$, we get

$$\rho_1 = \frac{kc(h^*-\theta)e^{-\frac{h^*-\theta}{k}}}{\theta(h^*-c)}.$$

On the other hand, if $c < \theta$, we can write the same equation into we can write the above equation into

$$h'(t) = G(t)e^{-ct},$$

where

$$G(t) = \frac{e^{-(\theta-c)t - \frac{h(t)-\theta}{k}[1-e^{-kt}]} - 1}{\frac{h(t)}{c(h(t)-\theta)}[1-e^{-ct}] + \frac{1}{h(t)-\theta}\left\{e^{-\theta t - \frac{h(t)-\theta}{k}[1-e^{-kt}]} - 1\right\}}.$$

Since it has been shown that $\lim_{t \to \infty} h(t) = h^* < c$, it is straightforward to show that $G(t)$ is uniformly bounded in t and

$$\lim_{t \to \infty} G(t) = \frac{kc(h^* - \theta)}{h^* - c}.$$

Now we postulate

$$h(t) \sim h^* + \rho_2 e^{-ct}, \quad \text{if } c < \theta.$$

Compare the limit of $h'(t)$, we get

$$\rho_2 = \frac{k(h^* - \theta)}{h^* - c}.$$

Corollary 6.3. *As $t \to \infty$,*

$$h(t) \sim h^* - \rho_1 e^{-\theta t} + \rho_2 e^{-ct},$$

where h^*, ρ_1, and ρ_2 are defined in Theorem 3.

6.6 Global Approximation Formulas

We propose that the free boundary $h(t)$ globally behaves like

$$h(t) \sim h^* - (h^* - c)e^{-\beta t}, \tag{6.13}$$

where clearly $h \to h^*$ as $t \to \infty$, and β is chosen to match the asymptotic expansion of $h(t) \sim c + \alpha t$, which means

$$\beta = \frac{k(c - \theta)}{3(h^* - c)} \tag{6.14}$$

The accuracy of approximation can be improved if we use a little bit more complicated interpolation formula

$$h(t) \sim h^* - (h^* - c)\exp[1 - e^{\beta t}]. \tag{6.15}$$

We choose a super exponential function to make the free boundary "decay" faster to the true boundary, when other conditions are matched as previous. And also this does not alter the asymptotic expansion at t infinity at all. In the same rationale, β is chosen to match the asymptotic expansion of $h(t) \sim c + \alpha t$ for t small, which gives the same expression of β as defined in (6.14). In Fig. 6.2 we provide a comparison of our analytical approximations and the true numerical solution of the free boundary.

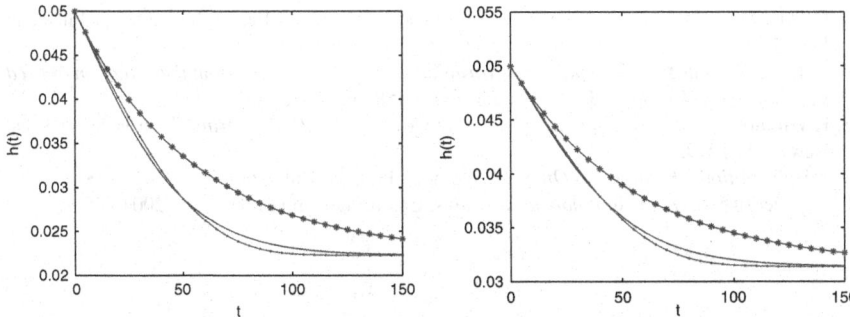

Fig. 6.2 The plain curve is the true solution. The top stared curve is the first approximation, and the bottom dotted curve is the second approximation. $c = 0.05, \theta = 0.06, k = 0.15$ (*left*), 0.10 (*right*). The units for t and $h(t)$ are years and year^{-1}, respectively

In general these approximation formulas are very accurate. Our numerical experiments with a variety of parameters show that the relative error is within 4% for $t < 20$ for the second formula. From the financial practitioner's point of view, both our numerical method and the approximation formula can provide satisfactory solutions.

6.7   Conclusion

Assuming the underlying interest rate follows the CIR model, we studied the mortgage borrower's optimal strategy to make prepayments when the volatility of market return rate is small. We derived the integral equation representation of the solution and studied the mathematical properties of the free boundary. An efficient iteration scheme was designed to solve the free boundary numerically. We also found two useful approximation formulas, the accuracy of which are validated with numerical simulations.

References

1. J. Cox, J. Ingersoll, & S. Ross, *A theory of the term structure of interest rates*, Econometrica , **53** (1985), 385–407.
2. Schwarts, E.S., & W.N. Torous, *Prepayment and the valuation of mortgage-backed securities*, J. of Finance **44**, 375–392.
3. D. Xie, X. Chen & J. Chadam, *Optimal Termination of Mortgages*, European Journal of Applied Mathematics, 2007.
4. L. Jiang & A. Rennie, *Mathematical Modeling and Methods of Option Pricing*, World Scientific Publishing House, 2005.
5. S.A. Buser, & P. H. Hendershott, *Pricing default-free fixed rate mortgages*, Housing Finance Rev. 3 (1984), 405–429.

6. D. Xie, *Parametric estimation for treasury bills*, International Research Journal of Financial Economics, 17 (2008), 27–32.
7. L. Jiang, B. Bian & F. Yi. *A parabolic variational inequality arising from the valuation of fixed rate mortgages*, European J. Appl. Math. **16** (2005), 361–338.
8. P. Willmott, *Derivatives, the theory and practice of financial engineering*, John Wiley & Sons, New York, 1999.
9. J. Hull, *Options, Futures and Other Derivatives* , Prentice Hall, 2005.
10. A. Etheridge, *A course in Financial Calculus*, Cambridge University Press, 2004.

Chapter 7
Analytical Design of Robust Multi-loop PI Controller for Multi-time Delay Processes

Truong Nguyen Luan Vu and Moonyong Lee

Abstract In this chapter, a robust design of multi-loop PI controller for multivariable processes in the presence of the multiplicative input uncertainty is presented. The method consists of two major steps: firstly, the analytical tuning rules of multi-loop PI controller are derived based on the direct synthesis and IMC-PID approach. Then, in the second step, the robust stability analysis is utilized for enhancing the robustness of proposed PI control systems. The most important feature of the proposed method is that the tradeoff between the robust stability and performance can be established by adjusting only one design parameter (i.e., the closed-loop time constant) via structured singular value synthesis. To verify the superiority of the proposed method, simulation studies have been conducted on a variety of the nominal processes and their plant-model mismatch cases. The results demonstrate that the proposed design method guarantees the robustness under the perturbation on each of the process parameters simultaneously.

Keywords Multi-loop PI controller · Direct synthesis · IMC-PID approach · Structured singular value

7.1 Introduction

Multi-loop (decentralized) PI control is still the vast majority of chemical process industry, despite a number of advanced control techniques can provide significant improvement such as the full multivariable controllers design methods. It is important to notice that the multi-loop PI controller has several merits for achievement, such as the simplicity in implementation, requiring fewer parameters to tune, and loop failure tolerance of control system can be assured during the design stage. Therefore, many controller tuning methods have been suggested to tune the

T.N.L. Vu and M. Lee (✉)
School of Chemical Engineering & Technology, Yeungnam University, 214-1, Dae-dong, Gyeongsan, Gyeongbuk 712-749, Korea
e-mail: mynlee@yu.ac.kr

S.-I. Ao et al. (eds.), *Advances in Machine Learning and Data Analysis*,
Lecture Notes in Electrical Engineering 48, DOI 10.1007/978-90-481-3177-8_7,
© Springer Science+Business Media B.V. 2010

95

multi-loop control system, which can be classified into four main types of tuning methods as follows: the detuning method, the independent design method, the sequential design method, and the relay auto-tuning.

In the detuning method, the individual controllers of the multi-loop control system are firstly designed by ignoring the interactions between the control loops, and then, all settings are detuned for taking into account the interactions until some stability criteria are satisfied. The well-known method of this type is the biggest log-modulus tuning (BLT) method suggested by Luyben [1]. Initially, the single-input, single-output (SISO) controllers are obtained by using the Ziegler-Nichols (Z-N) settings [2], the detuning is performed by adjusting one parameter F, wherein F is determined via a Nyquist-like plot of the closed-loop characteristic polynomial. The controller parameters for multi-loop control system are determined in such a way that the biggest log-modulus, which measures how far the control system is from being unstable, is obtained equal to $2N$ (N is the order of system) by adjusting F.

In the independent design method, each controller is designed based on the corresponding diagonal element of multivariable process, while the off-diagonal interactions should be taken into account by considering some inequality constraints on the process interactions. Therefore, many constraints imposed on the individual design are given by a number of authors (Grosdidier and Morari [3], Skogestad and Morari [4], etc.). The main advantage of independent design is that the failure tolerance is guaranteed automatically. However, it is conservative due to the assumption of the design method. This design method is effective when the system is diagonally dominance.

In the sequential design method [5, 6], each controller is designed sequentially with the previously designed controllers implemented. Basically, a controller is firstly designed by considering the selection of input–output pair and this loop is closed, and then a second controller is designed by considering the second pairing while the first loop is closed and so on. The sequential design method can be used for the complex interactive problems where the independent design method does not work. A potential disadvantage of this design method is that failure tolerance is not guaranteed when the previous loops fail. When the system outputs can be decoupled in time, the sequential design method can be effectively used for the design of multi-loop controllers.

Another widely used approach is the relay auto-tuning [7–9]. This approach is straightforward, because it directly combines single-loop relay auto-tuning and sequential tuning, wherein the multi-loop control system is tuned sequentially loop by loop, closing the ith loop when it has been tuned and the jth loop needs to be opened [9]. However, poor output responses can be obtained when the multiple-input, multiple-output (MIMO) system has large multiple time delays, which is one of the main causes of strong dynamic interactions.

To find a simple and effective design method of a multi-loop PI type controller with significant performance improvement has become an imperative research issue for process control engineers. In this chapter, a simple but efficient design method for multi-loop PI controllers is presented, which exploits the process interactions for the improvement of the loop performance. The proposed method is based on the direct synthesis [10, 11], wherein the multi-loop PI controller is designed based on the desired closed-loop transfer function via IMC approach [12–18]. In addition, the resulting analytical design rule includes a frequency-dependent relative gain array (RGA) [19–21] that provides information on the dynamic interactions useful for estimating the controller parameters. To secure robustness of the proposed PI control systems, the sufficient and necessary constraints for tuning the robust multi-loop controller are analyzed in presence of the process uncertainties. The feasibility of proposed method is verified by several examples.

7.2   The Multi-loop Feedback Controller Design for Desired Set-Point Responses

The conventional multi-loop feedback control system is shown in Fig. 7.1 where $G(s)$ represents the process transfer function matrix which is assumed that the process has a known model in the form of a square and non-singular transfer function matrix. $\tilde{G}_c(s)$ denotes the multiloop controller. $y(s)$ and $r(s)$ are the controlled variable and set-point vectors, respectively. The closed-loop response to the set-point change is obtained by

$$y(s) = H(s)r(s) = \left[I + \tilde{G}_c(s)G(s) \right]^{-1} \tilde{G}_c(s)G(s)r(s) \qquad (7.1)$$

where $H(s)$ is the closed-loop transfer function matrix.

The feedback controller to give a desired closed-loop response can be straightforwardly found by rearranging Equation (7.1). However, the resulting controller is generally not a diagonal (or decentralized) form. Instead of designing the multi-loop controller with n diagonal components directly, it is essential to consider only the diagonal matrix $H(s)$. Whereupon the multi-loop controller that results in $\tilde{H}(s)$ having n desired diagonal elements can be straightforwardly derived from Equation (7.1) by using some linear algebra, as shown in the following procedure:

$$\tilde{H}(s) = diag\left[\left(I + G(s)\tilde{G}_C(s) \right)^{-1} G(s)\tilde{G}_C(s) \right]$$
$$= diag\left[\left(G^{-1}(s)\tilde{G}_C^{-1}(s) + I \right)^{-1} \right] \qquad (7.2)$$

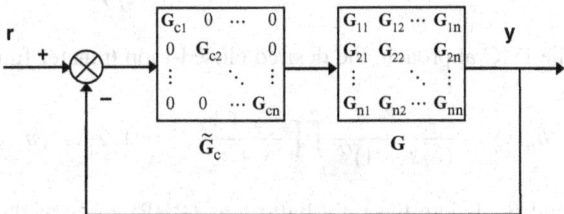

Fig. 7.1 Block diagram of multi-loop feedback control

The previous published papers [22, 23] demonstrate that the overall closed-loop system $\mathbf{H}(s)$ is stable if column diagonal dominance is achieved for all loops at all frequencies. Basically, in order to achieve the closed-loop stability, $\left(\mathbf{G}^{-1}(s)\tilde{\mathbf{G}}_C^{-1}(s) + \mathbf{I}\right)^{-1}$ is required to be closed-loop diagonal dominance or closed-loop column diagonal dominance at all frequencies. In particular, since $\tilde{\mathbf{G}}_C(s)$ has integral terms, this is the case for low frequency range. Thus, the inverse of matrix can be reasonably approximated as $\tilde{\mathbf{H}}^{-1}(s) \cong diag\left(\mathbf{G}^{-1}(s)\tilde{\mathbf{G}}_C^{-1}(s) + \mathbf{I}\right)$. Then one can obtain a multi-loop (or decentralized) feedback controller as follows:

$$\tilde{\mathbf{G}}_C(s) = diag\left(\mathbf{G}^{-1}(s)\right)\left(\tilde{\mathbf{H}}^{-1}(s) - \mathbf{I}\right)^{-1} \tag{7.3}$$

It is clear that the controller by Equation (7.3) gives a closed-loop response closer to the desired one as process interactions are insignificant. Since the multi-loop controllers are usually applied to processes with modest interactions, this approach can have validity.

To further simplify the above result, let $\mathbf{G}^{ij}(s)$ be the cofactor corresponding to $g_{ij}(s)$ in $\mathbf{G}(s)$. It follows from linear algebra that $diag(\mathbf{G}^{-1}(s)) = \frac{\mathbf{G}^{ii}(s)}{|G(s)|}$, where $|G(s)|$ denotes the determinant of $\mathbf{G}(s)$. Furthermore, $(\tilde{\mathbf{H}}^{-1}(s) - \mathbf{I})^{-1}$ can be expressed in terms of the ith diagonal element as $[\tilde{H}^{-1}(s) - I]^{-1} = diag\left\{\frac{h_{ii}(s)}{1 - h_{ii}(s)}\right\}$, of which $h_{ii}(s)$ is each diagonal element of $\tilde{\mathbf{H}}(s)$ and corresponds to the desired closed-loop transfer function for each loop. As a result, each diagonal feedback controller $g_{ci}(s)$ can be designed for g_{ii} according to Equation (7.3) as

$$g_{ci}(s) = \frac{G^{ii}(s)}{|G(s)|}\left(\frac{h_{ii}(s)}{1 - h_{ii}(s)}\right) \tag{7.4}$$

It follows that the diagonal element of the frequency-dependent RGA for $\mathbf{G}(s)$ is given by

$$\Lambda_{ii}(s) = g_{ii}(s)\frac{G^{ii}(s)}{|G(s)|} \tag{7.5}$$

Thus, each element of the multi-loop controller can be obtained as

$$g_{ci}(s) = \Lambda_{ii}(s)g_{ii}^{-1}(s)\left(\frac{h_{ii}(s)}{1 - h_{ii}(s)}\right) \tag{7.6}$$

According to the IMC approach, the desired closed-loop transfer function $h_{ii}(s)$ is chosen as

$$h_{ii}(s) = \frac{e^{-\theta_{ii}s}}{(\lambda_i s + 1)^{r_i}}\prod_{k=1}^{q_i}\frac{-s + z_k}{s + z_k^*}, i = 1, 2, \ldots, n \tag{7.7}$$

where z_k, z_k^* and θ_{ii} denote the right-haft-plane (RHP) zeros of the (i, i)th diagonal element of the process transfer function matrix, the corresponding complex conjugate of RHP zeros, and the time delay term, respectively q_i denotes the number of z_k. The IMC filter time constant, λ_i, which is also equivalent to the closed-loop time constant, is an adjustable parameter allowing an adequate tradeoff to be achieved between the performance and robustness. r_i is the relative order of the numerator and denominator in $g_{ii}(s)$.

Substituting Equation (7.7) into Equation (7.6), the multi-loop controller of the ith loop can be rewritten by

$$g_{ci}(s) = \Lambda_{ii}(s)g_{ii}^{-1}(s)\left(\frac{e^{-\theta_{ii}s}\prod_{k=1}^{q_i}\frac{z_k-s}{z_k^*+s}}{(\lambda_i s + 1)^{r_i} - e^{-\theta_{ii}s}\prod_{k=1}^{q_i}\frac{z_k-s}{z_k^*+s}}\right) \tag{7.8}$$

Note that in Equation (7.8), the non-minimum portion of $g_{ii}(s)$ is cancelled out with the time delay and RHP zero z_k in the numerator so that the controller has neither causality nor stability problem.

7.3   Reduction to the Multi-loop PI Controller

The multi-loop controller by Equation (7.8) is complicated and unaccepted in practice. Therefore, this controller is commonly transformed into the more practicable PI controller, which is one of the most acceptance controllers in the process industry. For the sake of simplicity, the mathematical Maclaurin series expansion is used to obtain the proposed multi-loop PI controller using the following procedure:

$$g_{ci}(s) \equiv s^{-1}p_{ii}(s) \tag{7.9}$$

Thus,

$$p_{ii}(s) \equiv s\Lambda_{ii}(s)g_{ii}^{-1}(s)\left(\frac{e^{-\theta_{ii}s}\prod_{k=1}^{q_i}\frac{z_k-s}{z_k^*+s}}{(\lambda_i s + 1)^{r_i} - e^{-\theta_{ii}s}\prod_{k=1}^{q_i}\frac{z_k-s}{z_k^*+s}}\right) \tag{7.10}$$

Consequently, Equation (7.9) can be expanded by using the Maclaurin series as

$$g_{ci}(s) = \frac{1}{s}\left[p_{ii}(0) + sp_{ii}'(0) + \ldots\right] \tag{7.11}$$

Since the standard form of multi-loop PI controller is given by

$$\tilde{\mathbf{G}}_C(s) = \frac{1}{s}\left[\tilde{\mathbf{K}}_I + s\tilde{\mathbf{K}}_C\right] \tag{7.12}$$

The proposed multi-loop PI controller is found by the comparison between Equations (7.11) and (7.12).

$$\tilde{\mathbf{K}}_C = diag\left\{ p'_{ii}(0) \right\}$$ (7.13)

$$\tilde{\mathbf{K}}_I = diag\left\{ p_{ii}(0) \right\}$$ (7.14)

Apparently, the above PI controller formulas can be utilized to achieve the control gain and integral time constant as an analytical function of the process model parameters and the closed-loop time constant, λ. Therefore, the proposed multi-loop PI controller can be tuned by a detuning parameter, λ, which is utilized to enhance the robust performance of the overall control system.

7.4 Example of Two-Input, Two-Output (TITO) Case

The TITO multi-delay processes are very popular in the process industry. In this section, the TITO multi-delay processes with the first-order plus delay time (FOPDT) dynamics are considered. The multi-loop feedback controller can be obtained from Equation (7.8) as

$$g_{ci}(s) = \Lambda_{ii}(s)\frac{(T_{ii}s + 1)}{K_{ii}}\left(\frac{1}{(\lambda_i s + 1) - e^{-\theta_{ii}s}}\right)$$ (7.15)

where K_{ii} and T_{ii} denote the gain and time constant of $g_{ii}(s)$, respectively. The order of the IMC filter is selected as 1 for the controller to be proper.

The (i, i)th element of the frequency-dependent RGA is calculated by

$$\Lambda_{ii}(s) = \frac{1}{1 - \dfrac{K_{12}K_{21}}{K_{11}K_{22}}\dfrac{(T_{11}s + 1)(T_{22}s + 1)}{(T_{12}s + 1)(T_{21}s + 1)}e^{-\theta_{e_i}s}}$$ (7.16)

where the effective delay θ_e is defined by $\theta_e = \theta_{12} + \theta_{21} - \theta_{11} - \theta_{22}$.

The multi-loop feedback controller can be found by substituting Equation (7.16) into Equation (7.15), and an analytical tuning rule of the proposed multi-loop PI controller can be derived by using Equations (7.13) and (7.14) as

$$K_{Ci} = \frac{\Lambda_{ii}(0)\left(\theta_{ii}^2 + 2\Lambda_{ii}(0)(\lambda_i + \theta_{ii})[K_e(T_{ei} - \theta_e) + T_{ii}]\right)}{2K_{ii}(\lambda_i + \theta_{ii})^2}$$ (7.17)

$$\tau_{Ii} = \frac{K_{Ci}}{K_{ii}} = \frac{\theta_{ii}^2 + 2\Lambda_{ii}(0)(\lambda_i + \theta_{ii})[K_e(T_{ei} - \theta_e) + T_{ii}]}{2K_{ii}(\lambda_i + \theta_{ii})}$$ (7.18)

where K_e denotes the interaction quotient [24], of which $K_e = \frac{K_{12}K_{21}}{K_{11}K_{22}}$. The effective time constant T_{ei} is defined as $T_{ei} = T_{jj} - T_{ij} - T_{ji}$, $j \neq i$. $\Lambda_{ii}(0)$ corresponds to the diagonal element of the steady-state RGA [25].

7.5   Robust Stability Analysis

Robustness is one of the most important issues that should be considered in control design because the control system stability and performance should be affected by the presence of process uncertainties. Generally, there is no nominal model which is a perfect mathematical representation of the actual system and the difference between the actual system and its nominal model may lead to closed-loop instability. Here, the robust stability analysis is carried out by focussing on the process multiplicative input uncertainty which is one of the most commonly used models to represent uncertainty in practice. Since the detailed analysis of robustness is mainly dependent on the uncertainty model. In this chapter, the structure singular value, μ, suggested by Doyle et al. [26] is considered as a measure of robustness of multi-loop feedback system.

As shown in Fig. 7.2, the process multiplicative input uncertainty can be interpreted as the process input actuator uncertainties and the actual process family may be described as $\Pi_I : G_p = G[I + W_I \Delta_I]$. Note that Π_I denotes the set of input perturbed process models. G_p is the $n \times n$ transfer function matrix of process model as perturbed from its nominal process model G due to the unstructured uncertainty as multiplicative input. The bound (weight) W_I are frequency-dependent and normalize the maximum magnitude of input perturbation block Δ_I to unity, where Δ_I is assumed to be stable with $\bar{\sigma}(\Delta_I) \leq 1$, $\forall \omega$. M_I involves all other blocks such as plant, controller, and weighting factors.

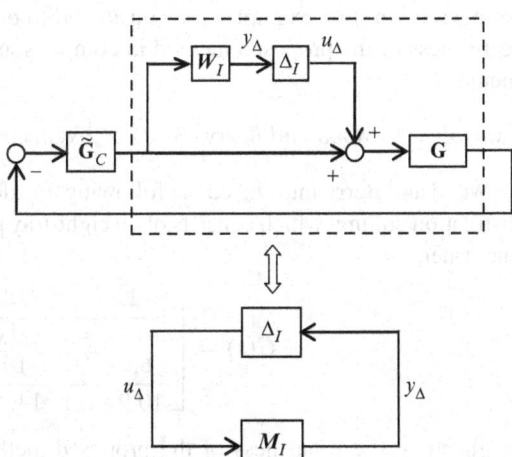

Fig. 7.2 Block diagram for multi-loop feedback system with multiplicative input uncertainty and its M-Δ control structure

Any perturbation blocks can be reorganized into the standard $M - \Delta$ control structure for robustness analysis, the transfer function matrix from the outputs to the inputs of Δ_I can be figured out as

$$M_I = -\left(I + \tilde{G}_c G\right)^{-1} \tilde{G}_c G W_I \tag{7.19}$$

It should be noted that M_I hold stability when the nominal control system has been conducted stable. On the basic of the Nyquist stability criterion, the overall control system is stable if and only if $det(I - M_I \Delta_I) \neq 0$, $\forall \omega$. Consider the structured singular value synthesis [21, 26–28], the multi-loop control system will remain stable under the multiplication input uncertainty if the following inequality constraint is satisfied:

$$\mu\left(M_I\left(j\omega\right)\right) = \mu\left(\left(I + \tilde{G}_c G\right)^{-1} \tilde{G}_c G W_I\right) < 1, \forall \omega \tag{7.20}$$

In this case, the weight W_I is used to normalize the maximum value of largest singular value of Δ_I at all frequencies.

Remarks:

1. $\mu(M_I) = 0$. In this situation, there does not exist any input perturbation block in multi-loop control system.
2. $\mu(M_I) = 1$. It is indicated that there exists a input perturbation block with $\bar{\sigma}(\Delta_I) = 1$, which is just large enough to make $I - M_I \Delta_I$ singular.
3. A smaller value of $\mu(M_I)$ is good. Inversely, a larger value of $\mu(M_I)$ is bad because it means that a smaller perturbation makes $I - M_I \Delta_I$ singular.

7.6   Simulation Study

In this section, two examples are considered to demonstrate the flexibility and effectiveness of the proposed method in comparison with those of other well-known methods.

Example 7.1. Wood and Berry (WB, 2 × 2 systems) [29] distillation column.

Wood and Berry introduced the following transfer function model of a pilot-scale distillation column which consists of an eight-tray plus reboiler separating methanol and water,

$$G(s) = \begin{bmatrix} \dfrac{12.8e^{-s}}{16.7s + 1} & \dfrac{-18.9e^{-3s}}{21s + 1} \\ \dfrac{6.6e^{-7s}}{10.9s + 1} & \dfrac{-19.4e^{-3s}}{14.4s + 1} \end{bmatrix} \tag{7.21}$$

To illustrate the robustness of the proposed method, assume there actually exist the diagonal multiplicative uncertainty in each manipulated input of magnitude $W_I(s) = diag\{(s + 0.3)/(s + 1), (s + 0.3)/(s + 1)\}$. This implies a relative uncertainty of up to 30% in the low frequency range, which increases with up to 100% uncertainty at high frequency range at about 1 rad/min. The uncertainty is represented as multiplicative input uncertainty as shown in Fig. 7.2.

Based on the μ-synthesis for multiplicative input uncertainty (Equation (7.20)), the set of the adjustable parameters λ_i $(i = 1, 2, \ldots, n)$ are suggested to achieve a desirable specification of robust stability and performance by increasing it monotonously. It can be seen in Fig. 7.3 that with a small λ_i of the multi-loop PI controller matrix $\tilde{G}_c$, the overall performances of proposed control systems can improve in terms of small IAE values and fast output responses, but the μ-values grow larger, which means that a smaller perturbation makes $I - M_I \Delta_I$ singular. On the contrary, increasing λ_i will slow down the overall output responses, but the μ-values grow smaller, tending to surpass their robust capacities in practice. Therefore, tuning the adjustable parameters λ_i provides a tradeoff between the achievable responses of robust performance and the robust stability. Consequently, to deal with the process uncertainties, it is suggested to increase the adjustable parameters λ_i monotonously, so that the nominal system responses will be gradually slow down for better robust stability. However, when the process un-modeled dynamics are too large, the robust performance and stability will not be acceptable. Then, the effective way to achieve better nominal system performance and robust stability is that reduce the process uncertainties.

For this example, it is desirable for the proposed controllers to be tuned by setting the closed-loop time constant λ_i as 1.3 and 7.3 for loops 1 and 2, respectively, so that the proposed control system can be guaranteed the robust stability $(\mu(M_I(j\omega)) = 0.80)$ as shown in Fig. 7.3. All of the controller parameters and the

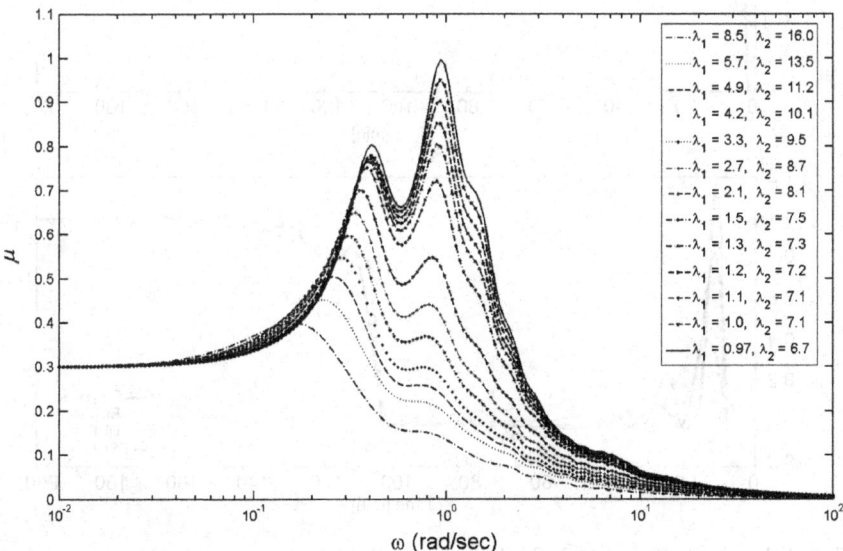

Fig. 7.3 μ-plots for WB column

resulting integral absolute error (IAE) values are tabulated in Table 7.1. The nominal responses of the closed-loops that are tuned by the proposed, BLT [1], DLT [30], and SAT [7] design methods are shown in Fig. 7.4. Simulation is carried out by introducing the unit set-point change sequentially at the first and second loops. It is apparent that the proposed PI control system provides the well-balanced and faster nominal responses in comparison with the other methods. Besides, the superiority of the proposed method is also illustrated by its IAE values in Table 7.1.

Table 7.1 PI parameters and IAE values for WB column

	Proposed	BLT	DLT	SAT
K_c	0.69, −0.08	0.38, −0.08	0.22, −0.14	0.87, −0.09
τ_I	10.05, 7.97	3.25, 23.60	17.20, 15.90	8.29, 10.40
IAE$_1$	3.41, 6.35	5.11,16.80	7.34, 5.64	4.00, 6.67
IAE$_2$	2.25, 10.59	3.38, 32.70	6.59, 10.91	0.81, 11.83
IAE$_t$	22.60	57.99	30.48	23.31

IAE$_i$: IAE for the step change in loop i.
IAE$_t$: sum of each IAE$_i$.

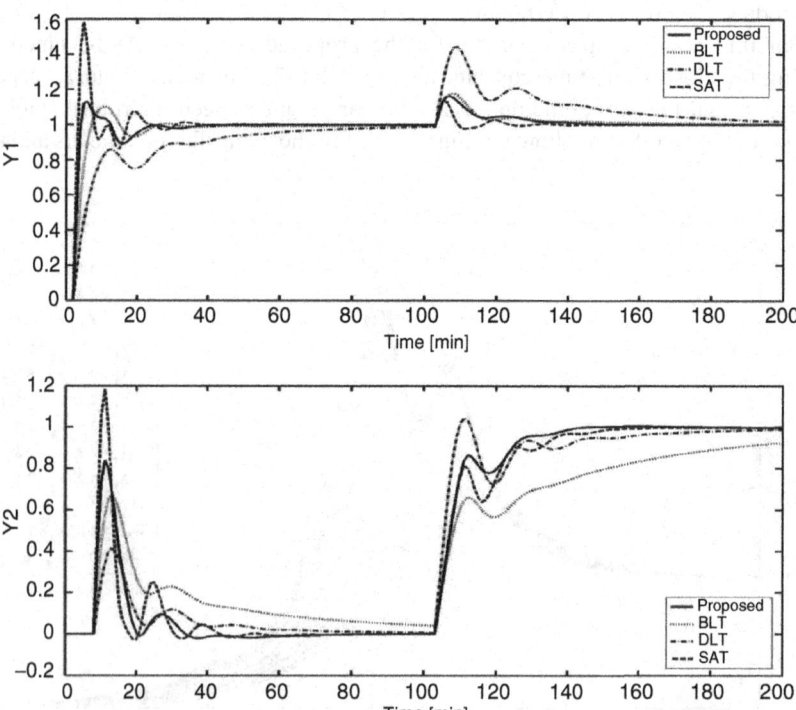

Fig. 7.4 Nominal responses for WB column

From the basis of the closed-loop responses in Fig. 7.4 and the IAE values in Table 7.1, one can see that the PI controller obtained by the proposed method provides better performance than those obtained by using the other methods.

Example 7.2. Alatiqi case 2 (A2, 4 × 4 system) [1] distillation column.

Consider the industrial distillation column studied by Luyben [1], the process transfer function matrix is given by

$$G(s) = \begin{bmatrix} \dfrac{4.09e^{-1.3s}}{(33s+1)(8.3s+1)} & \dfrac{-6.3e^{-0.2s}}{(31.6s+1)(20s+1)} & \dfrac{-0.25e^{-0.4s}}{(21s+1)} & \dfrac{-0.49e^{-5s}}{(22s+1)^2} \\[3mm] \dfrac{-4.17e^{-4s}}{(45s+1)} & \dfrac{6.93e^{-1.01s}}{(44.6s+1)} & \dfrac{-0.05e^{-5s}}{(34.5s+1)^2} & \dfrac{1.53e^{-2.8s}}{(48s+1)} \\[3mm] \dfrac{-1.73e^{-17s}}{(13s+1)^2} & \dfrac{5.11e^{-11s}}{(13.3s+1)^2} & \dfrac{4.61e^{1.02s}}{(18.5s+1)} & \dfrac{-5.48e^{-0.5s}}{(15s+1)} \\[3mm] \dfrac{-11.18e^{-2.6s}}{(43s+1)(6.5s+1)} & \dfrac{14.04e^{-0.02s}}{(45s+1)(10s+1)} & \dfrac{-0.1e^{-0.05s}}{(31.6s+1)(5s+1)} & \dfrac{4.49e^{-0.6s}}{(48s+1)(6.3s+1)} \end{bmatrix}$$

(7.22)

To demonstrate the robust stability of the proposed control system against the process multiplicative input uncertainties, assume that there exist the process diagonal multiplicative uncertainties in each manipulated input of magnitude $W_I(s) = diag\{(s + 0.1)/(s + 1), (s + 0.1)/(s + 1), (s + 0.1)/(s + 1), (s + 0.1)/(s + 1)\}$. This implies all of the process input actuators have up to 100% uncertainty at high frequency range at about 1 rad/min and almost 10% uncertainty in the low frequency range. By using the μ-synthesis (Equation (7.20)), the λ_i values can be found as 8, 2, 2, and 10 for loops 1, 2, 3, and 4, respectively, so that the proposed control system can be guaranteed the robust stability ($\mu(M_I(j\omega)) = 0.97$). All the controller parameters used in this example are listed in Table 7.2. To test the performance of control system, the unit step changes in set-point were sequentially made in the individual loops.

Figure 7.5 shows the nominal responses by the proposed and BLT methods. As shown in the figure, the proposed method provides the well-balanced and faster responses than the BLT method. Superiority of the proposed method can also be confirmed by its IAE values as shown in Table 7.2.

Table 7.2 PI parameters and IAE values for A2 column

	Proposed	BLT
K_c	2.22, 3.21, 2.88, 3.04	0.92, 1.16, 0.73, 2.17
τ_I	27.17, 14.31, 25.91, 169.7	61.7, 13.2, 13.2, 40
IAE_1	13.47, 4.05, 26.70, 83.25	50.96, 4.24, 43.43, 27.86
IAE_2	16.47, 4.74, 36.08, 73.13	80.97, 9.28, 100.20, 47.34
IAE_3	0.72, 0.2, 3.95, 5.3	3.25, 0.19, 6.68, 1.76
IAE_4	6.56, 1.72, 22.3, 18.69	21.64, 3.26, 53.17, 24.65
IAE_t	317.33	478.88

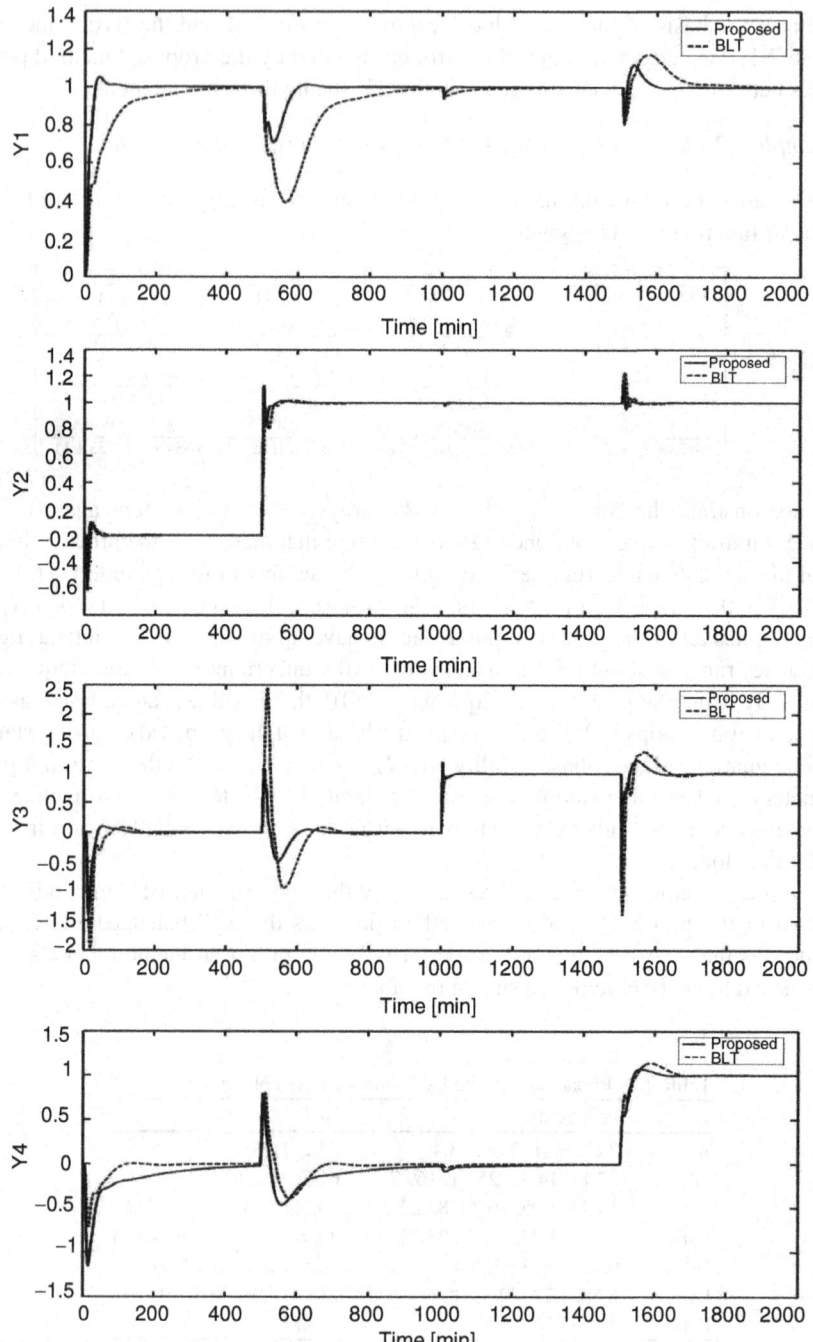

Fig. 7.5 Nominal responses for A2 column

7.7 Conclusions

A simple analytical design method for a robust multi-loop PI controller was proposed based on the direct synthesis and IMC-PID approaches for the multi-time delay processes. The proposed analytical tuning rules can take the interaction effect into account in the simple but efficient manner. By using the μ-synthesis, the proposed PI control systems can be afforded an excellent performance while the robustness of the multi-loop systems can be guaranteed satisfactorily. The simulations were conducted by tuning various multi-loop PI controllers for the diagonally dominant processes. The simulation results confirm the superiority of the proposed PI controllers over to several well-known existing methods.

References

1. W. L. Luyben, Simple method for tuning SISO controllers in multivariable systems, *Ind. Eng. Chem. Process Des. Dev.* **25**, 654–660 (1986).
2. J. B. Ziegler and N. B. Nichols, Optimum settings for automatic controllers, *ASME Tran.* **64**, 759–768 (1942).
3. P. Grosdidier and M. Morari, A computer-aided methodology for the design of decentralized controllers, *Comput. Chem. Eng.* **11**, 423–433 (1987).
4. S. Skogestad and M. Morari, Robust performance of decentralized control systems by independent design, *Automatica* **25**, 119–125 (1989).
5. D. Q. Mayne, The design of linear multivariable systems, *Automatica* **9**, 201–207 (1973).
6. M. Hovd and S. Skogestad, Sequential design of decentralized controllers, *Automatica* **30**, 1601–1607 (1994).
7. A. P. Loh C. C. Hang C. K. Quek, and V. U. Vasnani, Autotuning of multi-loop proportional – integral controllers using relay feedback, *Ind. Eng. Chem. Res.* **32**, 1102–1107 (1993).
8. S. H. Shen and C. C. Yu, Use of relay-feedback test for automatic tuning of multivariable systems, *AIChE J.* **40**, 627–646 (1994).
9. Y. Halevi Z. J. Palmor, and T. Efrati Automatic tuning of decentralized PID controllers for MIMO processes, *J. Process Control.* **7**, 119–128 (1997).
10. J. G. Truxal *Automatic Feedback Control System Synthesis* (McGraw-Hill, New York, 1955).
11. D. E. Seborg T. F. Edgar and D. A. Mellichamp, *Process Dynamics and Control* (Wiley New York, 1989).
12. C. E. Garcia and M. Morari, Internal model control I. A unifying review and some new results, *Ind. Eng. Chem. Proc. Process. Des. Dev.* **21**, 308–318 (1982).
13. D. E. Rivera, M. Morari, and S. Skogestad, Internal model control. 4. PID controller design: Multi-loop design. *Ind. Eng. Process Design. Dev.* **25**, 252–262 (1986).
14. C. G. Economou and M Morari, Internal model control 6: multi-loop design. *Ind. Eng. Chem. Proc. Process. Des. Dev.* **25**, 411–419 (1989).
15. M. Lee, K. Lee, C. Kim, and J. Lee, Analytical design of multi-loop PID controllers for desired closed-loop responses, *AIChE J.* **50**, 1631–1635 (2004).
16. N. L. V. Truong, J. Lee, and M. Lee, Analytical design of multi-loop PI controller for disturbance rejection in multivariable processes, *JCASE* **11**(5), 505–508 (2006).
17. N. L. V. Truong, J. Lee, and M. Lee, Design of multi-loop PID controllers based on the generalized IMC-PID method with *Mp* criterion, *IJCAS* **5**, 212–217 (2007).
18. N. L. V. Truong and M. Lee, Multi-loop PI controllers design for enhanced disturbance rejection in multi-delay processes, *IJMCS* **2**, 89–94 (2008).

19. M. E. Witcher and T. J. McAvoy, Interacting control systems: steady state and dynamic measurement of interaction. *ISA Trans.* **16**, 35–41 (1977).
20. E. H. Bristol, Recent Results on Interactions in Multivariable Process Control, in: AIChE Annual Meeting, Florida, 1978, p. 78b.
21. S. Skogestad and I. Postlethwaithe, *Multivariable Feedback Control Analysis and Design* (Wiley, New York, 1996).
22. P. Grosdidier and M. Morari, Interaction measures for systems under decentralized control. *Automatica* **22**, 309–319 (1986).
23. D. Chen and D. E. Seborg, Design of decentralized PI control systems based on Nyquist stability analysis, *J. Process Control.* **13**, 27–39 (2003).
24. J. E. Rijnsdorp, Interaction in two-variable control system for distillation columns-I and II. *Automatica* **1**, 15–28 (1965).
25. E. H. Bristol, On a measure of interactions for multivariable process control, *IEEE trans. Auto. Control.* **AC-11**, 133–134 (1966).
26. J. C. Doyle, J. E. Wall, and G. Stein, Performance and Robustness Analysis for Structured Uncertainty, in: Proc. IEEE Conf. Decision Contr., Orlando, FL, 1982, pp. 629–636.
27. M. Morari and Z. Zafiriou, *Robust Process Control* (Prentice Hall, Englewood Cliffs, NJ, 1989).
28. S. L. William, *Control System Fundamentals* (CRC Press, USA, 1999).
29. R. K. Wood and M. W. Berry, Terminal composition control of binary distillation column, *Chem. Eng. Sci.* **28**, 1707–1717 (1973).
30. J. Jung, J. Y. Choi, and J. Lee, One-parameter method for a multi-loop control system design, *Ind. Eng. Chem. Res.* **38**, 1580–1588 (1999).

Chapter 8
Automatic and Semi-automatic Methods for the Detection of Quasars in Sky Surveys

Sio-Iong Ao

Abstract With the advances of the technologies for the sky surveys, massive amount of survey data become available. It would be very helpful for the automatic and semi-automatic methods in the classifications/detections of the astrophysical objects. In fact, for surveys of millions of objects, it may not be possible to detect the desired objects by expert inspection alone. Quasars are interesting astrophysical objects that have been recently discovered more comprehensively from the sky surveys. Automatic and semi-automatic methods have been proposed for the detection of the quasars from the massive data produced by the modern sky surveys. In this chapter, the first section describes about the existing automatic and semi-automatic methods for the comprehensive search of quasars. Secondly, it will be explored to see if some machine learning algorithms can automatically classify the light curves of the quasars against the very similar light curves of the other stars. For example in MACHO sky survey, the light curves of the Be stars are so similar with the quasar light curves that the previous algorithms and even the manual examination by experts cannot tell the difference between the light. Experimental results will also be shown for this exploratory work.

Keywords Automatic detection · Quasar · MACHO · Autocorrelation · Clustering

8.1 Introduction

Quasars, quasi-stellar radio sources, are powerfully energetic and distant galaxy with an active galactic nucleus. Most quasars are farther than 3 billion light-years away from the Earth, but they are so luminous enough to be still visible. Their luminosity was found to vary on a variety of time scales. As the quasars are all very old, the studies of quasars may help us better understand the early stage of the universe.

S.-I. Ao (✉)
Harvard School of Engineering and Applied Sciences, Harvard University, Cambridge, MA, USA
e-mail: siao@comlab.ox.ac.uk; siao@harvard.edu

S.-I. Ao et al. (eds.), *Advances in Machine Learning and Data Analysis,*
Lecture Notes in Electrical Engineering 48, DOI 10.1007/978-90-481-3177-8_8,
© Springer Science+Business Media B.V. 2010

The quasars can also serve as suitable reference frames against which to measure the proper motion of other astrophysical objects [16]. They may also be helpful in the absorption line studies of the interstellar medium in the sky region. It is interesting itself to investigate the physical mechanisms underlying quasars light variation. The following works are some of the recent works on the investigation of the quasars light variation, which is shown to be helpful for the automatic quasar selection.

8.1.1 Variability Properties of the Quasar Light Curves

In the work of Scholz et al. [29] on the combined variability-proper motion survey for QSOs on Schmidt plates, in order to search for objects with variability timescales longer than a few months, the mean structure functions for time-lags longer than 100 days were compared with that for time-lags shorter than 100 days. The ratio $RS_{100} = SF(\tau > 100)/SF(\tau < 100)$ serves as an indicator of long-term variability. The criterion $\log_{10}(RS_{100}) > 0.15$ was adopted, which yielded a strong reduction of the candidate sample while only a few QSOs would remain undetected. The final sample contained 168 QSO candidates with estimated success of over 40% and completeness about 90%. Here, the variability index is defined as:

$$I_{\mathrm{var},i}^{(B)} = \frac{1}{n_{pl}^i} \sum_{j=1}^{n_{pl}^i} \frac{\left| B_{ij} - \bar{B}_i \right|}{\sigma_j \left(B_{ij} \right)} \text{ with } \bar{B}_i = \frac{1}{n_{pl}^i} \sum_{j=1}^{n_{pl}^i} B_{ij}$$

where $\sigma_j(B_{ij})$ is the photometric random error of the j-th plate at the magnitude B measured for the object on this plate. The proper motion index was defined as:

$$I_{pm} = \frac{\mu_x^2 + \mu_y^2}{\sqrt{\mu_x^2 \sigma_{\mu_x}^2 + \mu_y^2 \sigma_{\mu_y}^2}}$$

where (μ_x, μ_y) are the two proper motion components and $(\sigma_{\mu_x}, \sigma_{\mu_y})$ are their errors.

Meusinger et al. [19] conducted a combined variability and proper motion (VPM) survey on the globular cluster M 3. In their spectroscopic follow-up observations of QSO candidates, 114 QSOs and 10 Sey1s were pre-estimated with 90 completeness limit of the survey. In the VPM survey, the spectral energy distribution of QSOs is not used as a criterion. It is found that these VPM QSOs do not significantly differ from those samples from colour selection or slitless spectroscopy. As the variability of flux densities and stationarity of positions are two fundamental properties of quasars, the VPM survey is based on indices for star-like image structure, positional stationarity, overall variability, and long-term variability measured on 57 B plates taken with the Tautenburg Schmidt plates between 1964 and 1994.

The proper motion index, I_{pm}, is the measured proper motion in units of the proper motion error. The overall variability index, I_{var}, is the deviation of the individual magnitudes about the mean magnitude, and is normalized by the average magnitude scatter for star-like objects in the same magnitude range. I_{ltvar} is the means of structure function analysis and is computed for all star-like objects with $I_{pm} < 4$ and $B < 20$. The selection thresholds for the indices were based on the statistics of the previously known 90 QSOs in the field. With the set of constraints: $I_{pm} < 4$, $I_{var} > 1.3$, and $I_{ltvar} > 1.4$, the pre-estimated values for the success rate and the completeness are 90% and 40% respectively. With the spectroscopic follow-up observations of 198 candidates, it is estimated that the completeness of the VPM QSO sample with $B \leq 19.7$ is 94%.

Berk et al. [5] investigated how quasar variability in the rest-frame optical/UV regime depended on rest-frame time lag, luminosity, rest wavelength, redshift, the presence of radio and X-ray emission, and the presence of broad absorption line systems. A sample of over 25,000 quasars from the SDSS was used. The imaging photometry was compared with three-band spectrophotometry obtained at later epochs spanning time lags up to about 2 years. The results show that time dependence of variability (the structure function) is fitted well with a single power law of an index $\gamma = 0.246 \pm 0.008$, on timescales from days to years. It is found that there exist an anticorrelation of variability amplitude with rest wavelength, and a strong anticorrelation of variability with quasar luminosity. It is also shown that there exists a significant positive correlation of variability amplitude redshift.

The long-term quasar variability has been studied in works like Vries et al. [34]. In Vries' work, a sample of 35,165 quasars from the SDSS Data Release 2, along with 6,413 additional quasars in the same area of the sky from (2dF) QSO Redshift Survey, was used. The authors used the structure function (SF) for the variability diagnostic. The results on the quasar SF showed that most the long-term variations are intrinsic to the quasar itself. The results showed that there was no upper bound to the preferred variability timescale (smaller than a few decades). Second, the magnitude of the quasar variability was found to be a clear function of wavelength. Third, high-luminosity quasars tended to vary less than low-luminosity quasars. The results showed that the quasar variability was intrinsic to the source and was caused by chromatic outbursts/flares. Currently, the model that best explained this observed behavior was based on accretion DIs. In the work, the SF function is defined as

$$S(\tau) = \left\{ \frac{1}{N(\tau)} \sum_{i<j} [m(i) - m(j)]^2 \right\}^{1/2}$$

with the summation over all the combinations of measurements for with $\tau = t_j - t_i$. Here, the $n(n-1)/2$ permutations are grouped into bins that contain at least 200 measurements. The SF value for each bin is the RMS of the magnitude permutations.

The SF similarities are quantified in terms of their offset distributions. The mean SF offset and its 1 σ uncertainty, $\overline{O}$ and $\Delta \overline{O}$, are defined as

$$\overline{O} = \frac{1}{N} \sum_i^N S_A(i) - S_B(i)$$

$$\Delta \overline{O} = \frac{\sigma}{\sqrt{N}} = \frac{1}{N} \left\{ \sum_i^N [S_A(i) - S_B(i) - \overline{O}]^2 \right\}^{1/2}$$

where the SF curves are labeled A and B, both with $N = 42$ bins here, and $N \approx N^{1/2}(N-1)^{1/2}$.

The time lags between the optical continuum bands for a sample of 42 quasars were studied in the work [3]. The interpolation cross-correlation function (*ICCF*) method [14] was applied to the quasar sample [17], which was monitored in two colors (B and R). The maximum of *ICCF*(τ) showed the time delay between the bands. It was found that most of the objects showed a delay in the red light curve behind the blue one (a positive lag). The lag was about +4 days on average (+3 for the median). The results were broadly consistent with the reprocessing model.

In the following Section 8.2, several existing automatic and semi-automatic methods for the detection of the quasars from the massive data are reviewed, along with brief summaries of their results and limitations. In Section 8.3, some machine learning algorithms are explored for the automatic classification of the light curves of the quasars against the very similar light curves of the other stars. In Section 8.4, experimental results will also be shown for this exploratory work.

8.2 Automatic and Semi-automatic Methods for Quasar Detection

8.2.1 Variability Selection Method for Quasar Candidate in MACHO

Geha et al. [16] reported that 47 spectroscopically-confirmed quasars were discovered behind the Magellanic Clouds via photometric variability in the MACHO database. Thirty-eight quasars are behind the Large Magellanic Clouds (LMC) and nine behind the Small Magellanic Clouds (SMC). The follow-up spectroscopic detection efficiency was 20%, with emission-line Be stars as the primary contaminants. The first 5.7 years of MACHO data were used for the selection of the quasar candidates. The final 7.5 year MACHO light curves were used in the light curve analysis of quasars. Previously nine Active Galactic Nuclei (AGN) in MACHO database were cross-identified with X-ray sources by Schmidtke et al. [26], and one AGN by

Blanco and Heathcote [6]. Two additional sources in the MACHO database were presented by Dobrzycki et al. [10] subsequent to candidate selection. The light curves for these sources served as a training set for the developing of the selection method. The results showed that a total of 259 candidate quasars were observed for spectroscopic confirmation. Thirty-eight quasars were confirmed behind the LMC and nine behind the SMC. The quasars cover the redshift interval $0.2 < z < 2.8$ and mean magnitudes $16.6 \leq \bar{V} \leq 20.1$. The primary contamination during spectroscopic follow-up were the Be/Ae stars. Of the 258 quasar candidates observed, 188 were the emission-line Be/Ae stars.

The variability selection method for quasar candidate, designed to automatically reject known classes of variable stars, with the final step being a selection by eye, was shown as followed:

Data: Deviation from a constant brightness light curve

There are 12 million objects in MACHO LMC and 2 million objects in MACHO SMC. From the Level 1 MACHO database, 140,000 objects were flagged as having a significant deviation from a constant brightness light curve [2], and the proposed selection method began with these objects. The objects were required to have a minimum of 50 photometric measurements in V- and R-bands.

Step 1: Magnitude and color tests

Weighted average magnitudes were calculated from the standard equation:

$$\bar{V} \equiv \sum_{i=1}^{N} \frac{V_i}{\sigma_{V,i}^2} \Big/ \sum_{i=1}^{N} \frac{1}{\sigma_{V,i}^2}$$

where N is the total number of individual photometric measurements V_i with corresponding errors $\sigma_{V,i}$. Candidates were required to be with $16 \leq \bar{V} \leq 20$ and have weighted average colors bluer than $(\overline{V - R}) \leq 1.0$. The color cut eliminated long-period quasi-periodic variable stars, while including all the AGNs in the training set.

Step 2: Statistics tests of light curve variability

Two statistics were used to check the amount of light curve variability. The intrinsic variability was as:

$$\widehat{\sigma_V} \equiv \sqrt{\frac{\sum (V_i - \bar{V})^2}{N-1} - \frac{\sum \sigma_{V,i}^2}{N}}$$

It is required that the intrinsic variability of MACHO candidate quasars to be larger than 0.05 magnitudes ($\widehat{\sigma_V} \geq 0.05$), as the sample of 42 quasars by Giveon et al. [17] found $0.05 \leq \widehat{\sigma_B} \leq 0.32$ in the B-band. The second variability statistics is the variability index [35]:

$$I \equiv \sqrt{\frac{1}{N(N-1)} \sum_{i=1}^{N} \left(\frac{V_i - \bar{V}}{\sigma_{V,i}}\right) \left(\frac{R_i - \bar{R}}{\sigma_{R,i}}\right)}$$

The index measures the correlated variability between the MACHO V- and R-bands, and approaches zero for uncorrelated variability. As quasar variability is expected to be highly correlated between the MACHO passbands, the selection criterion is set as $I \geq 1.5$ for quasar candidates.

Step 3: Removal of periodic variable stars

The periodic variable stars, such as Cepheid and RR Lyrae stars, were removed using the MACHO Variable Catalogue. Light curves of a period (τ) at high significance in both passbands over the range $0.1 \leq \tau \leq 500$ days were rejected.

Step 4: Examination by eye

Candidate light curves were examined by eye for removing objects with spurious noise characteristics or quasi-periodic components. About 2,500 light curves were examined by eye, with a total of 360 light curves considered candidate quasars.

In addition, the authors have search for optically-variable counterparts in the error boxes of suspected extragalactic radio and X-ray sources in several Magellanic Cloud surveys.

8.2.2 Variability Selection Methods for Quasar Candidate in OGLE-II

Eyer [12] proposed variability selection method to find QSO (quasi-stellar object, or quasar) candidates behind the Small and Large Magellanic Clouds (SMC and LMC). The optical variability properties of the points sources from the OGLE-II photometric data. The LMC catalog has 7 million stars, and SMC data has 2.2 million stars. The variable star catalog listed 53,000 stars for LMC and 15,000 stars for SMC. The overall results were that 118 and 15 QSO candidates are identified for the LMC and SMC, respectively, but, the confirmation with spectroscopic data has not been done yet. The schematic description of the selection criteria was as followed.

First criterion: selection on photometry: magnitude and colour

The motivation for the selection on photometry is that the color-magnitude for the stars observed by OGLE-II and for the variable stars has different distributions. Objects bluer than $V - I = 0.9$ mag and within 17 mag $< I < 20.5$ mag were selected. This is to reject the time series of the red giant branch variables stars. From the experience with SDSS, it is estimated that a cut $I > 17$ mag will exclude one or two QSOs in all fields together. For the LMC data, this magnitude cut selected a population of main sequence blue variables, RR Lyr stars and red giant stars. Some short period Cepheids are also included in the SMC data. The results showed that 6,241 objects are selected for LMC and 1,553 objects for SMC.

Second criterion: the slope of variograms

It is known that the QSOs usually have little variability at short time scales. The variability is irregular and aperiodic and increases as longer time-scales are observed.

The variograms/structure functions are employed for the selection of variability for time scales longer than 100 days. The variograms functions are computed as followed: All pair differences of time, $h_{ij} = JD_j - JD_i$, and squared pair differences of magnitude, $(I_j - I_i)^2$, are calculated, where h is the lag and JD is the Julian Date. The median of the subsample, $2\Upsilon_{med}(h)$, of $(I_j - I_i)^2$ formed by all possible pairs i, j, where $h_{ij} < h$. The variogram can estimate the spread of the distribution formed by the $I_j - I_i$. Its slope can serve as a discriminating criterion to select the time series that has increasing variability for longer time scales.

In Eyer's work, the objects with a compute slope of variogram less than 0.1 for $h \geq 100$ days are rejected. The authors established the value for this limit empirically (with the variable QSO, behind LMC, OGLE050833.29-685427.5), and also by conducting Monte Carlo simulations on the SMC data.

Third criterion: QSO and Be star colors

With this slope selection criterion, the list of QSO candidates reduced to 649 objects for LMC and 179 for SMC. The selection cut at $B - V = 0.04$ mag can reject 87% of Be stars and reject less than 1% of the QSOs. This selection on color can cut the sample by half to 312 stars.

Forth criterion: Manual selection

A few undesirable effects, like spurious monotonic slopes, field overlap, effect of a bleeding column or extended wings, brighter or dimmer step-like variation in the data etc., were identified manually. Then the authors selected manually the 312 stars on individual basis thereby to eliminate these undesirable effects.

Follow-up spectroscopy experiments and works

Dobrzycki et al. [11] conducted the follow-up spectroscopic observations of quasars candidates in the SMC selected by Eyer [12] from the OGLE database. Of the 12 QSO candidates, 5 are confirmed quasars. Two of these quasars were also independently identified by Geha et al. [15]. The emission redshifts of the five quasars range from 0.27 to 2.17. It shows that Eyer's method is efficient and the contaminants in the list of candidates were predominantly early type stars in the SMC. No obvious trend or difference was observed between the light curves of these quasars and stars. The optical spectra were obtained with the Magellan Baade 6–5 m telescope. The spectra of all five quasars show at least two emission lines. This can enable the unambiguous determination of emission redshifts.

Dobrzycki et al. [9] presented their discovery of nine quasars behind the Large Magellanic Cloud. Six of them were identified as part of the systematic variability-based search for QSOs in the objects from the OGLE-II database. This variability-based identification was performed by Eyer [12], and the list formed the main part of the candidate pool. Several candidates from the X-ray selected sample (Dobrzycki et al. 2002) were also followed up. The remaining three quasars were identified by follow-up spectroscopy of optical counterparts to X-ray sources.

For the variability selected quasars, during the authors' observing runs, observations of 108 objects out of the remaining 111 candidates were completed. Six of them were identified as QSOs. For the X-ray selected quasars, a total of 30 objects were observed and three was found to be quasars. Combining with previous found quasars from the OGLE-II LMC and SMC databases, there are 14 quasars selected. From the colour–colour diagram, the authors found that setting a cutoff at $V - I \approx 0.3$ may increase the efficiency of the variability-based selection method, to 40–50%. It was also found that a decent correlation ($R = 0.64$) is seen for quasars.

Sumi et al. [31] selected 97 QSO candidates via variability from OGLE-II DIA variable star catalogue in the Galactic Bulge fields. Spectroscopic Follow-up observations were planned to confirm the QSO candidates in four of these fields (BUL_SC1, 2, 32 and 45). Sumi et al. used the VI photometric maps of OGLE-II fields [33], and the light curves of objects in the OGLE-II Galactic Bulge variable star catalogue [36]. 30,219 objects which are fainter than $I_0 = 16$ mag are selected, where I_0 is the I-band magnitude.

Following the procedure of Eyer [12], variogram function is used to find objects with increasing variability on longer time-scales. The slope of the variogram is defined as: $S_{\mathrm{var}} = dVar(h)/d \log h$, where differences of times, $h_{i,j} = t_j - t_i$. For a given bin of $\log h$, $Var(h)$ is the median of $(I_j - I_i)^2$, where $(I_j - I_i)^2$ is square of difference of magnitudes. The AoV algorithm [30] is employed to search the periodicity in the light curves. The light curves in which strong periodicity is detected with AoV periodogram of larger than 50 were rejected, and 3,201 light curves remain. The remaining light curves were inspected by eye. Firstly, the apparent periodic and semi-periodic variables were rejected. Then stars with contamination like proper motion objects and artifacts around bright Variable stars were rejected. Finally 97 QSO candidates were selected.

8.2.3 Automatic Photometric Selection of Quasars from the Sloan Digital Sky Survey

Sandage and Wyndham [25] pioneered the selection of quasars from multicolor imaging data and researches on this area continued through the years. The proposed selection algorithm seeks to explore all the regions of color space that quasars are known to occupy. In the work of Richards et al. [22], the quasar candidates were selected via their nonstellar colors in $ugriz$ broadband photometry and by matching unresolved sources to the FIRST radio catalogs. Nearly 95% of previously known quasars were recovered, and the overall completeness is expected to be over 90%. The overall efficiency is better than 65%. The SDSS quasar survey will be approximately four times the size of the concurrent 2dF QSO Redshift Survey [7]. The approach is similar to previous studies, but certain characteristics of the SDSS make the selection of SDSS quasars unique.

Briefly, the proposed algorithm works as follows: (1) Objects with spurious fluxes in the imaging data are rejected. (2) Point-source matches to FIRST radio sources [4] are preferentially targeted without reference to their colors. (3) The sources remaining after the first step are compared to the distribution of normal stars and galaxies in two distinct three-dimensional color spaces. Once quasar candidates were identified, the SDSS fiber-fed spectrographs were used to obtain the spectra of each candidate.

In the color selection process, the first step is to remove from consideration the region inhabited by stars. It is known that the colors of ordinary stars occupy a continuous, almost one-dimensional region in $(u-g)$, $(g-r)$, $(r-i)$, $(i-z)$ color–color–color–color space [20]. As a result of their distinct colors, quasars candidates can be identified as outliers from the stellar locus. The proposed algorithm chooses quasar candidate objects that are more than 4σ from the stellar locus, where σ can be determined from the errors of the objects in question and the width of the stellar locus at the nearest point. The selection is limited to objects fainter than $i^* = 15$. The $ugri$-selected objects are targeted to $i^* = 19.1$, while $griz$-selected objects are targeted to $i^* = 20.2$. Outliers from the stellar locus are targeted for follow-up spectroscopy.

The completeness of the proposed algorithm is estimated by checking with the recovering of the previously known quasars and the selection of the simulated quasars. A region of about $100\,\mathrm{deg}^2$ of sky that has been targeted by the final version of the algorithm has been used to test the efficiency of the algorithm. It is found that 1,113 of the 1,687 quasar candidates with spectra are quasars or AGNs, with an efficiency of 66.0%.

Follow-up Works to This Automatic Photometric Selection of Quasars

In the follow-up work of Richards et al. [23], a non-parametric Bayesian classification, based on kernel density estimation, was applied to parameterize the color distribution of astronomical sources. A catalog of 100,563 unresolved, UV-excess (UVX) quasar candidates to $g = 21$ from $2,099\,\mathrm{deg}^2$ of the Sloan Digital Sky Survey (SDSS) Data Release One (DR1) imaging data was presented. From the existing spectral of 22,737 sources, 97.6% were found to be quasars. It was estimated that 95.0% of the objects in the catalog are quasars. The Bayes' Rule was applied to weight each likelihood with its corresponding prior probability to compute the posterior probability of an object x being a star or quasar:

$$P\left(C_1|x\right) = \frac{p\left(x|C_1\right) P\left(C_1\right)}{p\left(x|C_1\right) P\left(C_1\right) + p\left(x|C_2\right) P\left(C_2\right)}$$

where C_1 and C_2 are the star classes. Here, objects with $P(C_1|x) > 0.5$ are classified as stars, and objects with $P(C_1|x) < 0.5$ are classified as quasars.

Nonparametric Bayesian Classification (NBC)

The quasar training set is the four primary SDSS colors $(u-g, g-r, r-i, i-z)$ of the 16,713 quasars from Schneider et al. [27]. The colors from a random sample of 10% of all point sources in DR1 imaging area with $14.5 < g < 21.0$ were also used. Those objects that do not pass the photometric quality tests [22] were rejected. The spectroscopically confirmed quasars were rejected too. The total number of objects in the initial training set was 478,144. The test set composed of SDSS-DR1 point sources with $u - g < 1.0$. Those objects that do not pass the photometric quality tests [22] were rejected. The full set contained 831,600 objects.

The likelihood of each object x in the test set with respect to each training set was computed with the nonparametric kernel density estimator:

$$\hat{p}(x) = \frac{1}{N} \sum_i^N K_h \left(\|x - x_i\|\right)$$

where N is the number of data points, $K_h(z)$, the kernel function, satisfied $\int_{-\infty}^{\infty} K_h(z)dz = 1$, h is a scaling factor called the bandwidth, and z is the "distance" between a pointing the test set to a point in the training set. Here, these distances are the 4-D Euclidean color differences, $\|x - x_i\|$, and a Gaussian kernel,

$$K_h \left(\|x - x_i\|\right) = \frac{1}{h\sqrt{2\pi}} \exp -\frac{\|x - x_i\|^2}{2h^2}$$

was used.

Fast Algorithms for Computing the Kernel Density Estimate

A naïve algorithm for computing the kernel density estimate at N points among N points would requires N^2 distance operations. Richards et al. used a fast computational algorithm based on space-partitioning trees and principles similar to those used in N-body solvers [18]. A second fast algorithm was proposed for quickly finding the higher posterior probability with details by Gray. Its advantage is that it need not estimate the density completely for each object to be classified. The algorithm only needs to maintain upper and lower bounds on the density for reach class. The bandwidth pair was determined with a statistical criterion for optimal classification accuracy, i.e. leave-one-out, cross-validated accuracy here. The proposed fast NBC algorithm can compute the leave-one-out accuracy score for each pair of bandwidths for the two training sets fast. The best bandwidth was found to be 0.15 mag for each training set, and the corresponding accuracies were 94.48% for the quasars and 97.91% for the stars.

The proposed NBC algorithm was applied to identify quasars. Further cleaning was performed by rejecting objects with large KDE stellar probabilities. This resulted in the 100,563 quasar candidates in the catalog. To estimate the completeness of the proposed approach, the authors checked the results with the SDSS-DRI quasars in the Schneider et al. [27] catalog. It was found that 94.7% of the quasars were recovered with the proposed algorithm. It was estimated that the filtering of objects via their photometric flags (e.g., those with "fatal errors") may have an effect of 5% additional incompleteness. Simulation approach was also employed to evaluate the completeness of the proposed algorithm. It estimated that the approach is generally at least 95% complete between $z = 0.2$ and $z = 2.0$.

To estimate the efficiency of the catalog, the catalog was matched with three spectroscopic databases, i.e. SDSS-DR1, 2QZ NGP catalog [8], and SDSS-DR2 [1]. In all there were 22,737 matches to spectroscopically confirmed objects. 97.6% were confirmed to be quasars. The authors expected that the overall efficiency of the catalog would be 95.0%, obtaining 95,502 quasars in all, even though it may have difficulties to extrapolate the efficiency for these confirmed objects to the entire sample. Further improvements for the algorithm may be achieved in the future with regard to high-redshift and extended quasars. The selection to fainter limits may also be extended. The inclusion of properties like radio- and X-ray detections and lack of proper motion into the algorithm may also improve the efficiency.

Another follow-up work is the one by Richards et al. [24]. The proposed Bayesian algorithm was applied to identify 1,172,157 quasars candidates from a sample of over 40 million SDSS point sources. The overall efficiency of the catalog is about 80%, while a UVX subsample with over 500,000 objects has an expected efficiency of over 97%. Cross-comparison with spectroscopically confirmed type 1 quasars in the COSMOS field was conducted. It was found that the sample is at least 70% complete. The photometric redshifts were estimated for the full sample with an expected accuracy of ±0.3 in about 80% of the time.

The photometric imaging data was from the SDSS Data Release 6. Point sources (type = 6) with i-band magnitudes between 14.5 and (de-reddened) 21.3 ($\text{psfmag}_i > 14.5$ and $\text{psfmag}_i\text{-extinction}_i < 21.3$) were used. The 77,429

SDSS-DR5 quasars with spectra [28] were used for the quasar training set. The set was further supplement with the AAOmega-UKIDSS-SDSS (AUS) QSO Survey, which added another 304 spectroscopically confirmed quasars. The quasars of the $z > 5.7$ by the SDSS [13] were also included. In addition, the training set included the 920 objects from the cross-comparison of SDSS and *Spitzer* data. The final quasar training set included 75,382 confirmed quasars. The full test set included 44,449,609 objects to be classified.

The Bayesian classification algorithm used in the paper is similar to the Paper I [23], with a few changes to the procedure. An improvement over the algorithm in Paper I is on the determination of the optimal bandwidth for classification. An initial broad search of possible bandwidths is now attempted, and then a narrower search around the most optimal bandwidth is executed. The final bandwidths were found 0.11 mag for stars and 0.12 for quasars, with the estimated accuracy (completeness) of 92.6% for the quasar training set. Here, the authors classified the objects as low-redshift ($z \leq 2.2$), mid-redshift ($2.2 < z < 3.5$) and high-redshift ($z \geq 3.5$). The motivation is that the distribution of quasar colors changes considerably with redshift.

8.3 Methodology for the Exploration of the Quasar Autocorrelation

8.3.1 Experimental Data and Data Processing

In Geha's work, 47 spectroscopically-confirmed quasars were discovered behind the Clouds. Along with the additional 12 quasar objects previously identified, a total of 59 quasars and 116 Be stars are studied here. Linear interpolation is applied to these light curves to obtain uniformly distributed time series datasets. The linear interpolation is a curving fitting method with linear polynomials. Given two points (x_0, y_0) and (x_1, y_1), the linear interpolant is the straight line between the points. For a point (x, y) in the interval (x_0, x_1), the y coordinate is given by:

$$y = y_0 + (x - x_0) \frac{y_1 - y_0}{x_1 - x_0}$$

In this work, the linear interpolation is used to estimate the observation values which are uniformly distributed in time, and it is observed that the time series with the linear interpolation are very similar in shape with the original time series. The uniform distribution property is helpful for the computing of the correlation of the light curves.

8.3.2   *Autocorrelation of the Time Series*

Then, the autocorrelation is employed to measure the similarity between observations between the observations of the processed light curve of the linear interpolation as a function of the time separation between. Repeating patterns of a light curve may be identified by the autocorrelation. The autocorrelation is estimated with:

$$\hat{R}(k) = \frac{1}{(n-k)\sigma^2} \sum_{t=1}^{n-k} [X_t - \mu][X_{t+k} - \mu]$$

for any positive integer $k < n$, where $X_1, X_2, \ldots X_n$ are the observations of the light curve, μ and σ are the estimated mean and variance of the light curve.

In the regression of the time series data, the existence of the autocorrelation violates an assumption of the ordinary least squares (OLS) that there is no correlation between the error terms. This violation can cause the standard errors to be estimated inaccurately. Besides the autocorrelation test, in this work, the coefficients of the constant terms and the first lag terms of the autoregression of the quasar light curves and the Be star light curves have also been compared.

The moving average is employed to obtain some long-term pattern of the light curves. The moving average is a series of averages of different subjects of the full time series dataset. A simple moving average is the mean of the previous n data points, while a central moving average uses both the past and future data to compute the averages. Given a sequence $\{a_i\}_{i=1}^N$, a central $(2n + 1)$-moving average is use, with a new sequence $\{s_i\}_{i=1}^N$ as followed:

$$s_i = \frac{1}{m_i} \sum_{j=i-n, where\ j>0\ and\ j<=N}^{i+n} a_j$$

where m_i is the number of terms in the summation of a_j from $(j = i - n)$ to $(j = i + n)$, where j is required to be larger than zero and not larger than N for any given value i. Here, the central five-moving average is used. The subtraction of the long-term signal from the original light curve signal can produce the short-term (high-frequency) components.

8.3.3   *Dynamic Time Warping*

Dynamic time warping (DTW) is used to measure the similarity between two time sequences. It can allow a non-linear mapping of one sequence to another by minimizing the distance between the two [21]. DTW method can be applied for various tasks in time series problems like clustering, classification and anomaly detection etc. An advantage of DTW is its flexibility to allow two sequences that are similar but locally out of phase to align in a non-linear way. The time complexity of DTW is $O(n^2)$. For the problems like this study of a few hundred sequences, each with a few hundred data points, the CPU cost of DTW is simply not an issue. Otherwise, speed improvements like smaller warping windows can be used to speed up the DTW calculations.

In this work, the pattern similarity between the autocorrelation sequences of the light curves is determined by aligning two autocorrelation sequences A and B, with distortion $D(A, B)$, which is two of the autocorrelation sequences of the light curves. The objective is to have a decision rule that can choose the alignment with the smallest distortion:

$$R^* = \arg \min D(A, B)$$

Dynamic time warping is employed to obtain the best possible alignment warp between A and B, along with the associated distortion R^*. Mathematically, let the two sequences be represented as $A = \{a_1, \ldots, a_N\}$ and $B = \{b_1, \ldots, b_M\}$. An alignment warp, $\varphi = (\varphi_a, \varphi_b)$, aligns A and B through a point-to-point mapping of length K_φ

$$a_{\varphi_a(k)} \Leftrightarrow b_{\varphi_b(k)} 1 \leq k \leq K_\varphi$$

The overall distortion $D_\varphi(A, B)$ of a particular alignment warp φ is the sum of the local distances between elements $d(a_{\varphi_a(i)}, b_{\varphi_b(j)})$:

$$D_\varphi(A, B) = \sum_{k=1}^{K_\varphi} d\left(a_{\varphi_a(k)}, b_{\varphi_b(k)}\right)$$

With the boundary constraints $\varphi_a(1) = \varphi_b(1) = 1$, $\varphi_a(K) = N$, $\varphi_b(K) = M$, and the monotonicity property $\varphi_a(k + 1) >= \varphi_a(k)$, $\varphi_b(k + 1) >= \varphi_b(k)$. The optimal alignment is the one that minimizes the overall distortion:

$$D(A, B) = \min_\varphi D_\varphi(A, B)$$

As an example, for two sequences of signal 1: [3 4 10 6 8] and signal 2: [9 10 12 13 10 17], the computed optical alignments are [3 4 10 10 10 10 6 8] and [9 9 10 12 13 10 10 17] respectively, as shown in the following figure (Fig. 8.1).

8.3.4 Hierarchical Clustering of the Autocorrelation Sequences

In this work, the hierarchical clustering is employed to cluster the above sequences of the autocorrelations of the quasar and Be star light curves. The distance functions are estimated with the DTW method. The goal of the clustering algorithms is to figure out the underlying similarities among a set of feature vectors x, and to cluster similar vectors together [32]. The clustering algorithms have many different applications in social sciences, engineering and medical science. The algorithms can

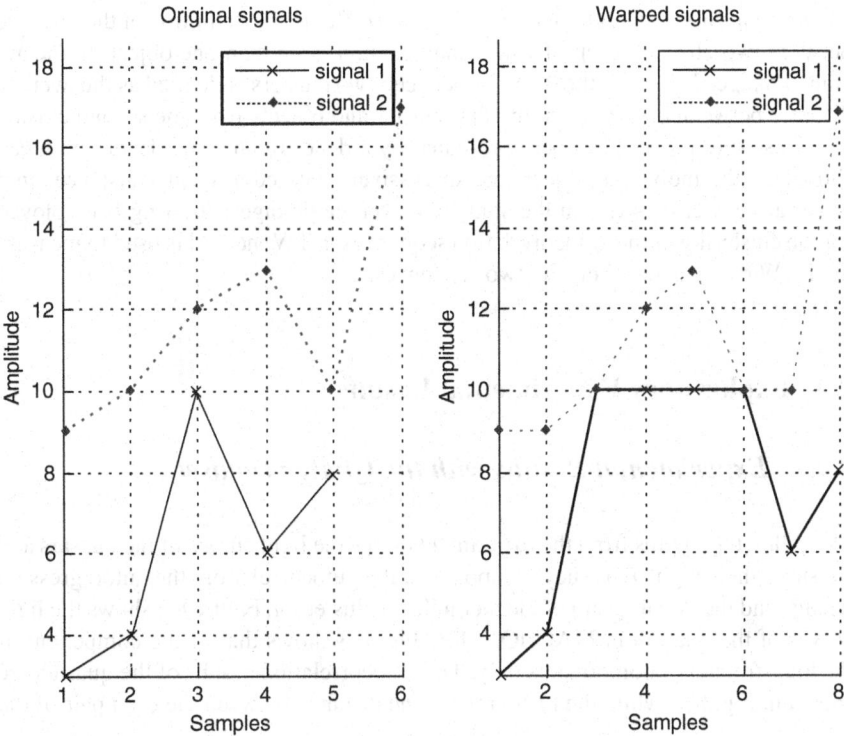

Fig. 8.1 The original signals (*left*) versus the warped signals by the dynamic time warping method (*right*)

provide us with clusters that are characterized by the similarity of vectors within each cluster. When a new data set or pattern is available, we can assign it to the known cluster by comparing its characters with each cluster's characters.

The clustering process can be viewed as a combinatorial problem of putting the data points into optimal clusters. However, it is NP-hard to enumerate all such possibilities of clustering. In view of the computational difficulty, different clustering algorithms have been developed so that only a small number of the different possible combinations of the clusters will be considered. A popular approach is the agglomerative clustering, where the two clusters with the smallest inter-cluster distance are successively merged until all the objects have been merged into a single cluster. It can be described with the concept of nesting. Let $\Re_1$ of k clusters and $\Re_2$ of r clusters be two clusters form in the clustering process, where $r < k$. When each cluster in $\Re_1$ is a subset of a set in $\Re_2$, then we say that $\Re_1$ is nested in $\Re_2$ and we denote this by $\Re_1 \subset \Re_2$. For example, let $\Re_1 = \{\{x_2, x_4\}, \{x_1\}, \{x_3, x_5\}\}$ and $\Re_2 = \{\{x_2, x_4\}, \{x_1, x_3, x_5\}\}$, then we say that $\Re_1$ is nested in $\Re_2$. The hierarchy of the agglomerative algorithms is as follow. Let the initial clustering be $\Re_0$, the first clustering be $\Re_1, \ldots$, and the final clustering be $\Re_{N-1}$. The hierarchy is $\Re_0 \subset \Re_1 \subset \ldots \subset \Re_{N-1}$.

Different forms of agglomerative clustering differ in the definition of the distance between two clusters, each of which may contain more than one object. In the average linkage clustering, the distance between two clusters is defined as the average distance between objects from the first cluster and objects from the second cluster. The clustering process can be represented by a dendrogram. The dendrogram can show how the individual objects are successively merged at greater distances into larger and fewer clusters. In the study, the average linkage clustering is employed for the clustering of the autocorrelation sequences. DTW method is used to measure the DTW distance between any two sequences.

8.4   Exploratory Experimental Results

8.4.1   Experimental Results with the Quasar Dataset

The following figures from the experiments show the light curves of the quasars and Be stars, their high-frequency components, the autocorrelation, the autoregression results, and the dendrogram of the hierarchical clustering. Figure 8.2 shows the light curves of the quasars in MACHO, while Fig. 8.3 shows that of the component of the high-frequency components only. The autocorrelation graphs of the quasars are shown in Fig. 8.4, while the histogram of the distance between the each pair of the

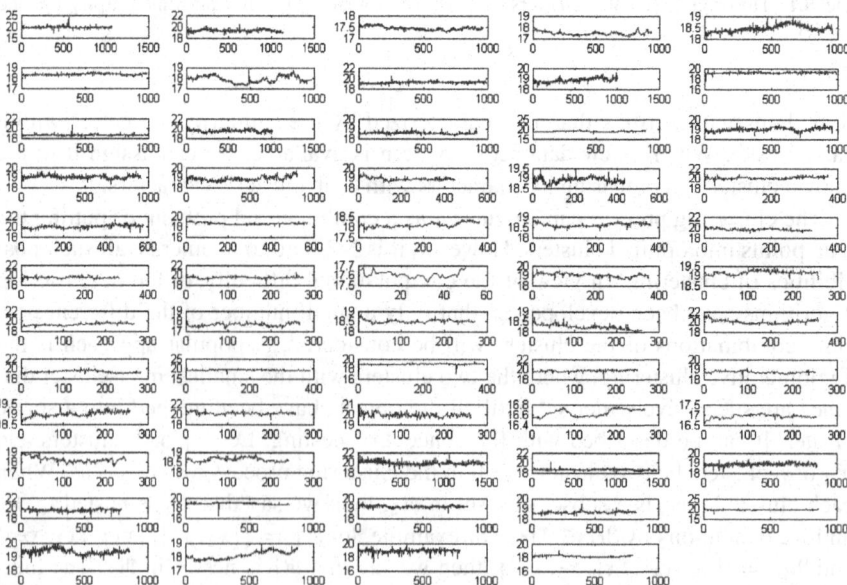

Fig. 8.2 The light curves of the quasars in MACHO (with the numbering of the quasars from left to right, then from top to bottom). The horizontal axis is the date of observation, while the vertical axis is the calibrated observed values of the light curves

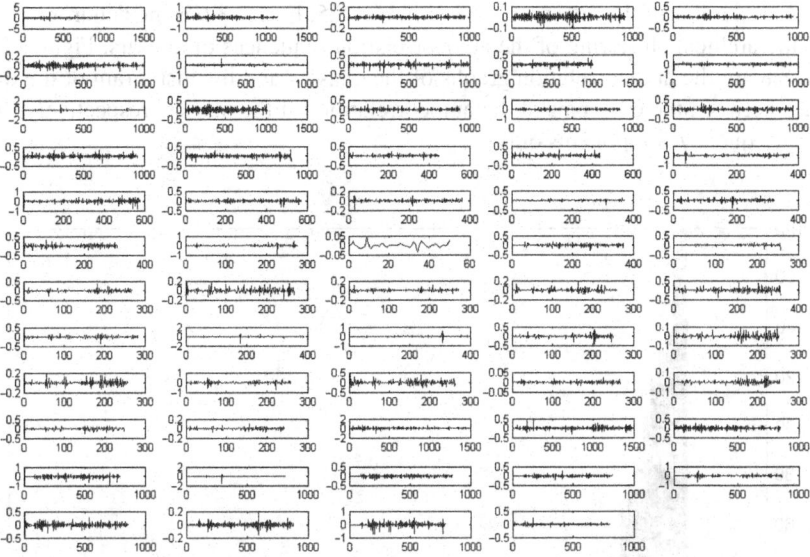

Fig. 8.3 The light curves of the quasars in MACHO, showing the component of the high-frequency components only. The horizontal axis is the date of observation, while the vertical axis is the calibrated observed values of the light curves

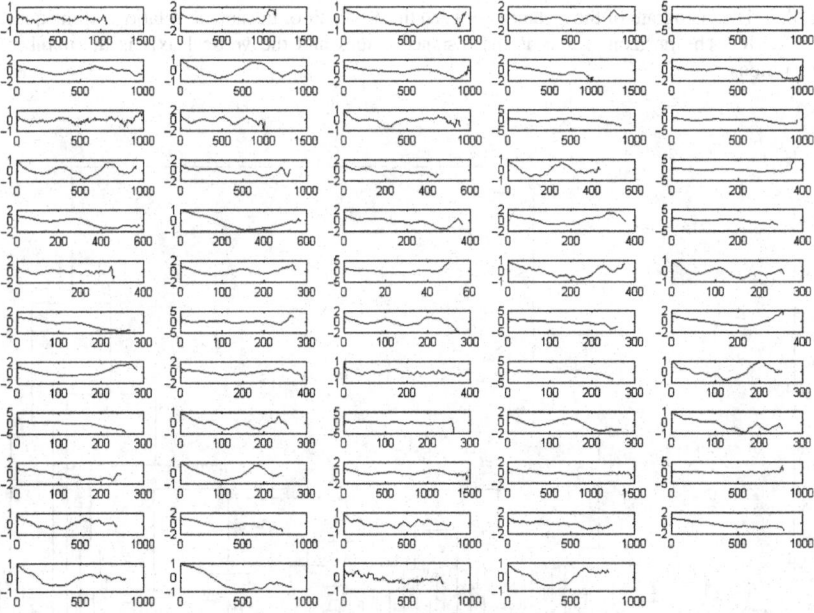

Fig. 8.4 The autocorrelation graphs of the quasars in MACHO. The horizontal axis is the date of observation, while the vertical axis is the computed correlation values of the light curves

autocorrelations of the quasars is shown in Fig. 8.5. Figure 8.6 is the dendrogram of
the hierarchical clustering of the autocorrelation sequences of quasars. Figures 8.7
and 8.8 are the autocorrelation graphs of the quasars and the histogram of the distance between the each pair of the autocorrelations of the quasars respectively (the
high-frequency components only).

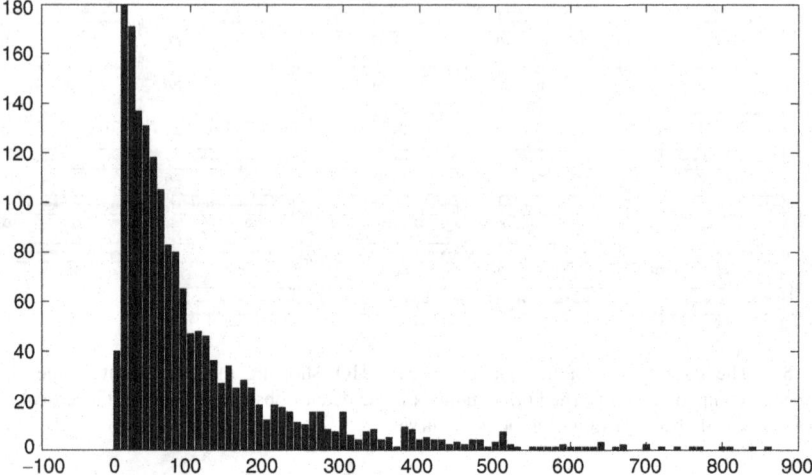

Fig. 8.5 The histogram of the distance between the each pair of the autocorrelations of the quasars
in MACHO. The horizontal axis is the distance values and the vertical axis is the number of
observations

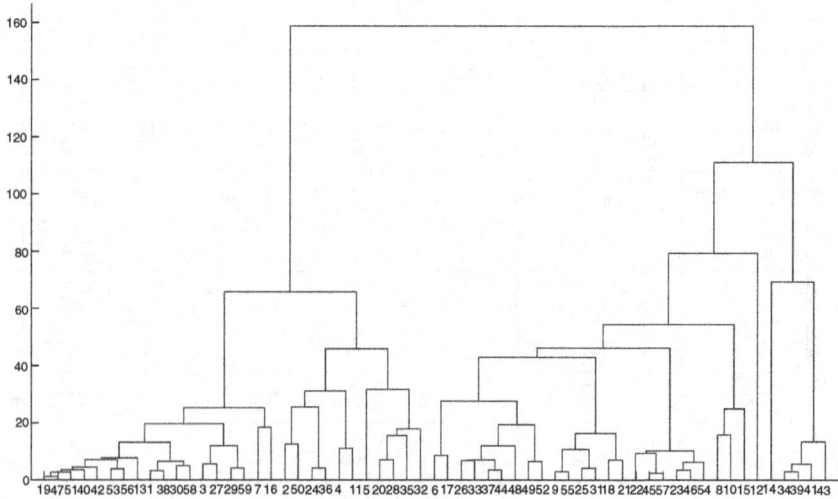

Fig. 8.6 The dendrogram of the hierarchical clustering of the autocorrelation sequences of quasars
in MACHO. The labels in the horizontal axis are the assigned quasar number, while the vertical
axis is the DTW distance between the quasars

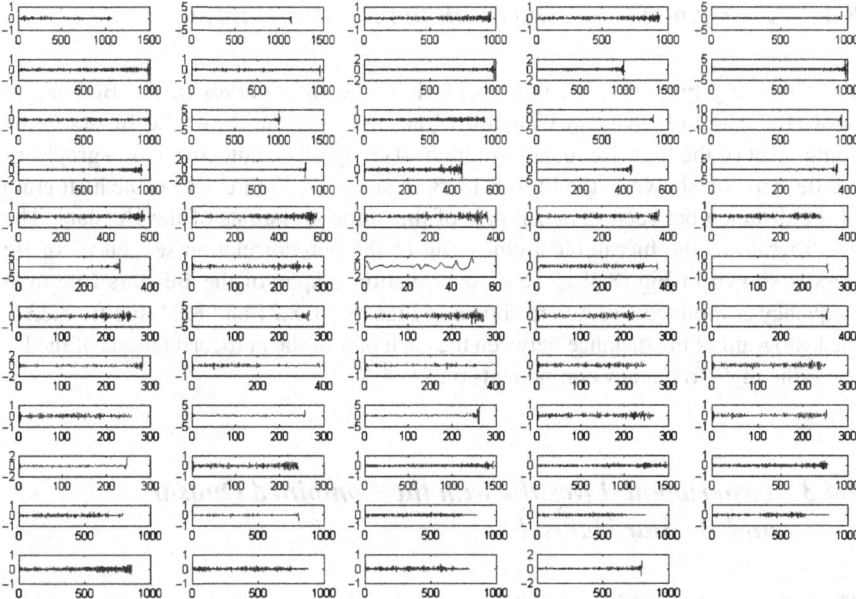

Fig. 8.7 The autocorrelation graphs of the quasars in MACHO (the high-frequency components only). The horizontal axis is the date of observation, while the vertical axis is the computed correlation values of the light curves

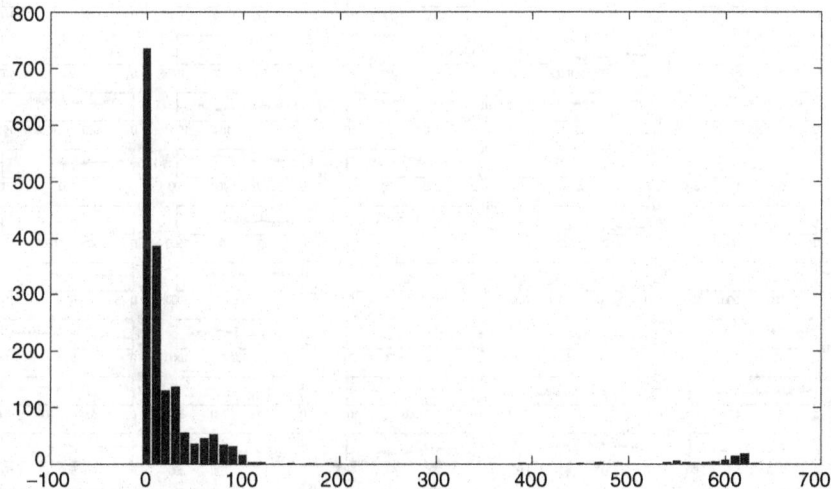

Fig. 8.8 The histogram of the distance between the each pair of the autocorrelations of the quasars in MACHO (the high-frequency components only)

8.4.2 Experimental Results with the Be Star Dataset

The following Figs. 8.9, 8.10 and 8.11 show the light curves of the Be stars in MACHO. The Figs. 8.12, 8.13 and 8.14 are the light curves of the Be stars (the component of the high-frequency components only). The autocorrelation graphs of the Be stars are shown in the Figs. 8.15, 8.16 and 8.17. Figure 8.18 is the histogram of the distance between the each pair of the autocorrelations of the Be stars. The dendrogram of the hierarchical clustering of the autocorrelation sequences of Be stars is shown in Fig. 8.19. The autocorrelation graphs of the Be stars (the high frequency components only) are shown in Figs. 8.20, 8.21 and 8.22. Figure 8.23 is the histogram of the distance between the each pair of the autocorrelations of the Be stars (the high-frequency components only).

8.4.3 Experimental Results with the Combined Quasar and Be Star Dataset

The following Figs. 8.24 and 8.25 are the plots of the coefficients of the constant terms versus those of the first lag terms of the autoregression equations of the quasars, and Be star respectively. It was noticed that the plots for the quasars and

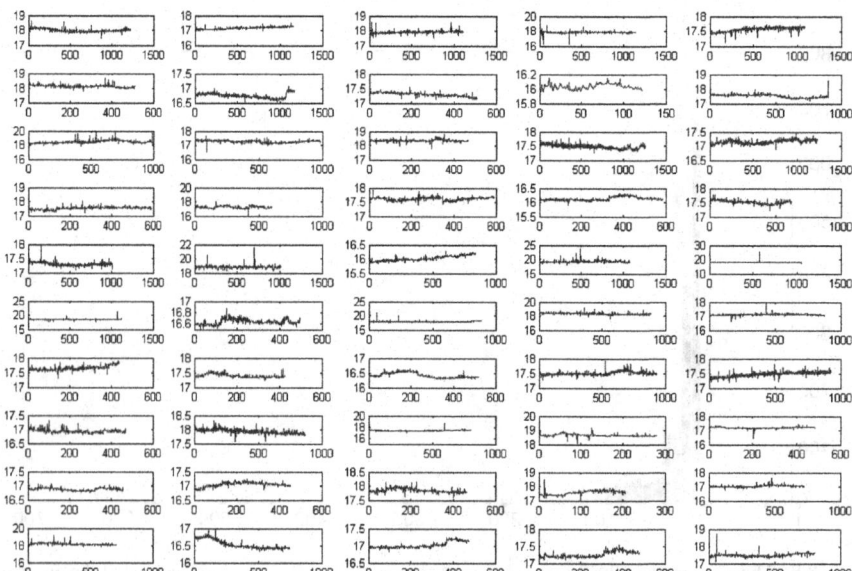

Fig. 8.9 The light curves of the Be stars (Part I) in MACHO (with the numbering of the quasars from left to right, then from top to bottom). The horizontal axis is the date of observation, while the vertical axis is the calibrated observed values of the light curves

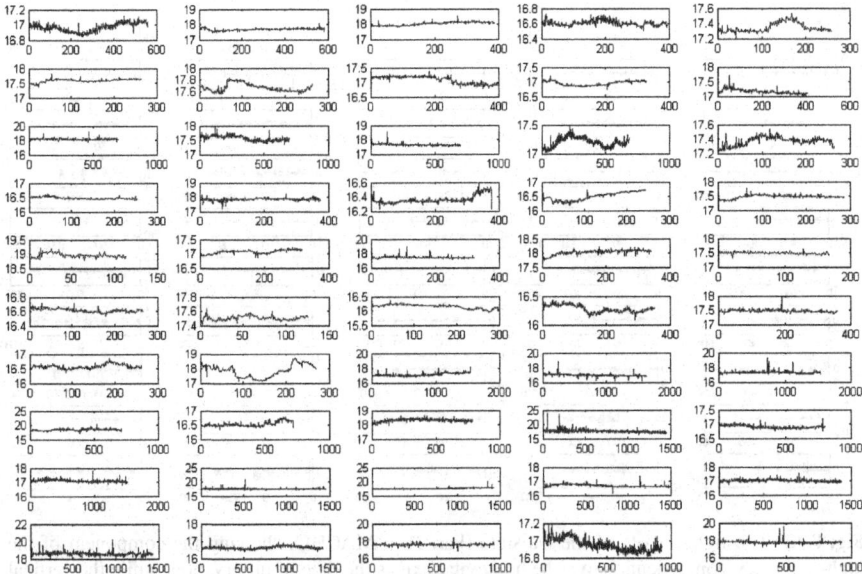

Fig. 8.10 The light curves of the Be stars (Part II) in MACHO (with the numbering of the quasars from left to right, then from top to bottom). The horizontal axis is the date of observation, while the vertical axis is the calibrated observed values of the light curves

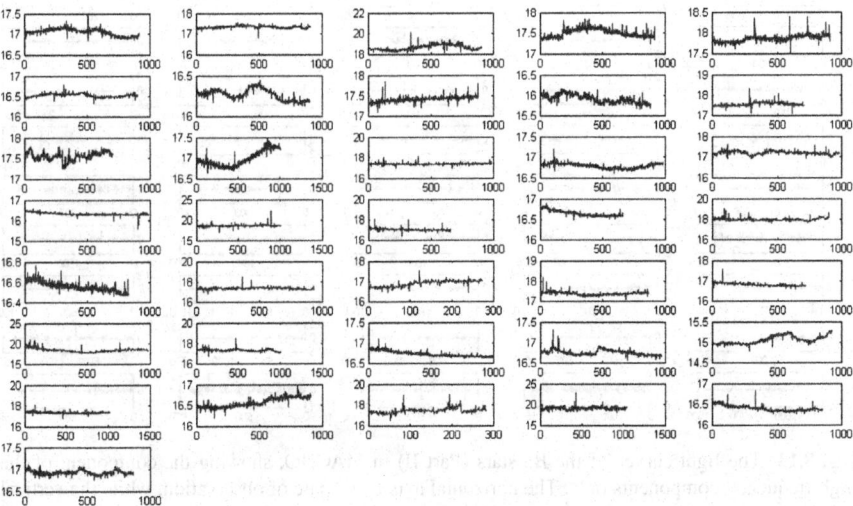

Fig. 8.11 The light curves of the Be stars (Part III) in MACHO (with the numbering of the quasars from left to right, then from top to bottom). The horizontal axis is the date of observation, while the vertical axis is the calibrated observed values of the light curves

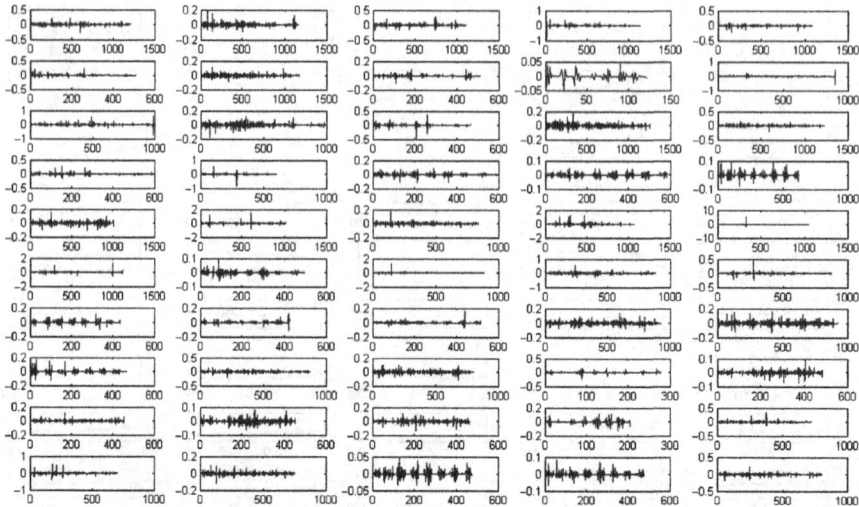

Fig. 8.12 The light curves of the Be stars (Part I) in MACHO, showing the component of the high-frequency components only. The horizontal axis is the date of observation, while the vertical axis is the calibrated observed values of the light curves

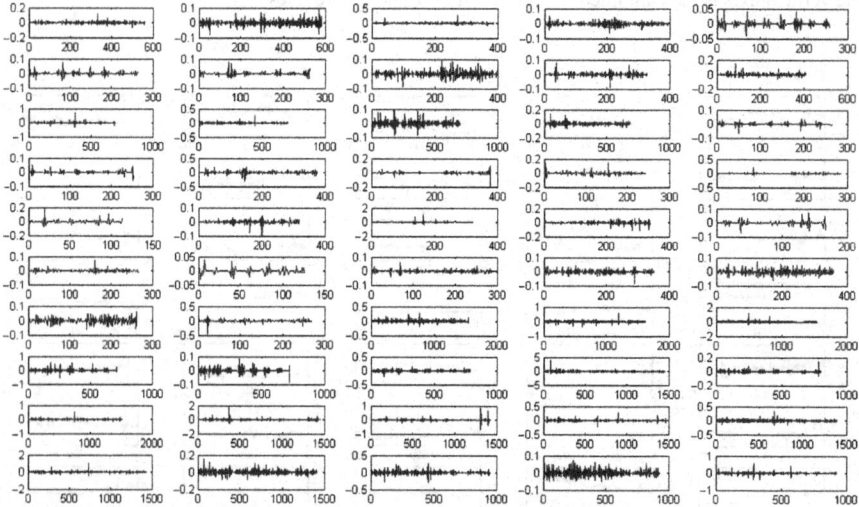

Fig. 8.13 The light curves of the Be stars (Part II) in MACHO, showing the component of the high-frequency components only. The horizontal axis is the date of observation, while the vertical axis is the calibrated observed values of the light curves

Be stars are very similar with each other. In Fig. 8.26, the hierarchical clustering of the autocorrelation sequences of the quasars and Be stars are shown. It can be easily observed that it produced two large-size clusters and one very small outlier cluster.

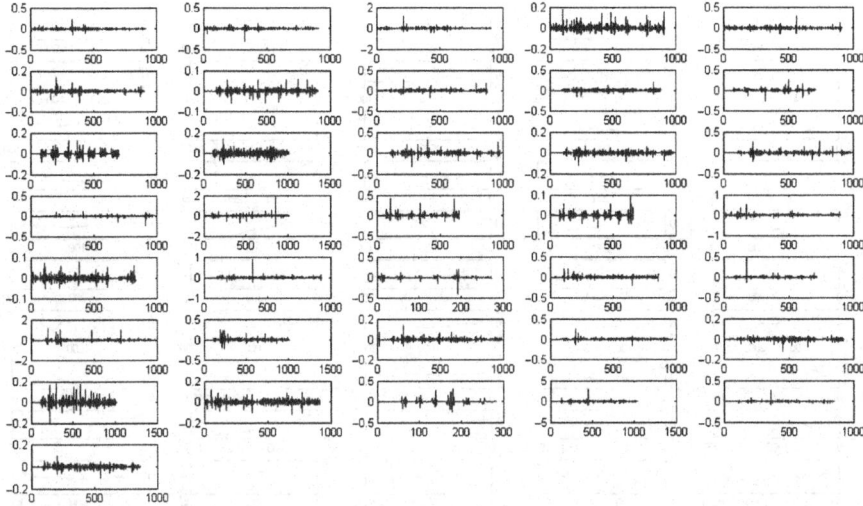

Fig. 8.14 The light curves of the Be stars (Part III) in MACHO, showing the component of the high-frequency components only. The horizontal axis is the date of observation, while the vertical axis is the calibrated observed values of the light curves

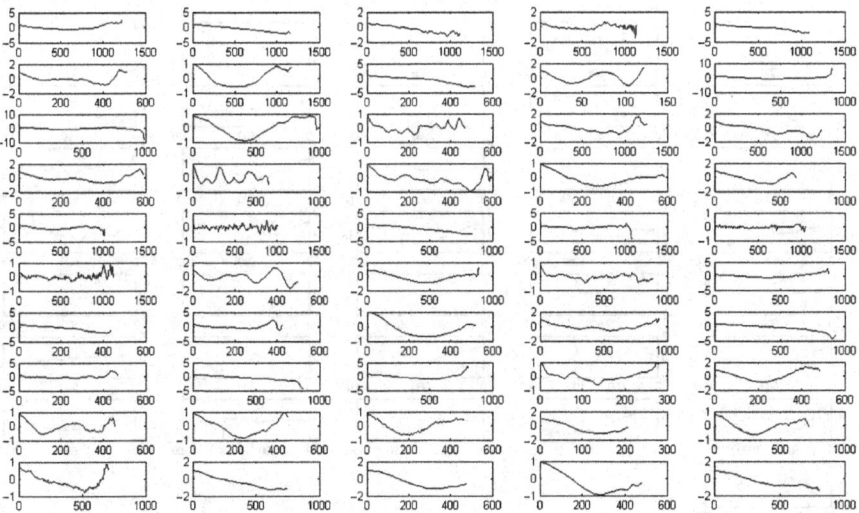

Fig. 8.15 The autocorrelation graphs of the Be stars (Part I) in MACHO. The horizontal axis is the date of observation, while the vertical axis is the computed correlation values of the light curves

The small outlier cluster consists of three Be stars. The first large-size cluster consists of 29 quasars and 50 Be stars, while the second one consists of 30 quasars and 83 Be stars. This seems to suggest that the autocorrelation functions of the quasars and Be stars cannot be easily distinguished with each other.

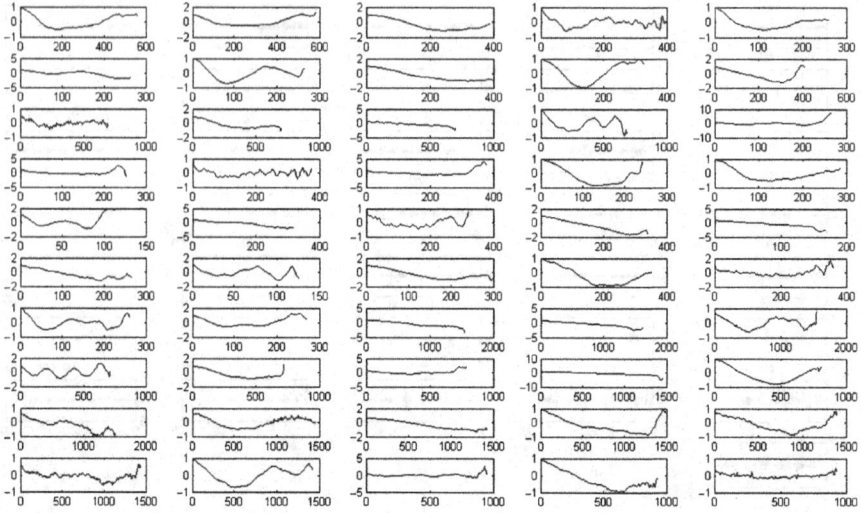

Fig. 8.16 The autocorrelation graphs of the Be stars (Part II) in MACHO. The horizontal axis is the date of observation, while the vertical axis is the computed correlation values of the light curves

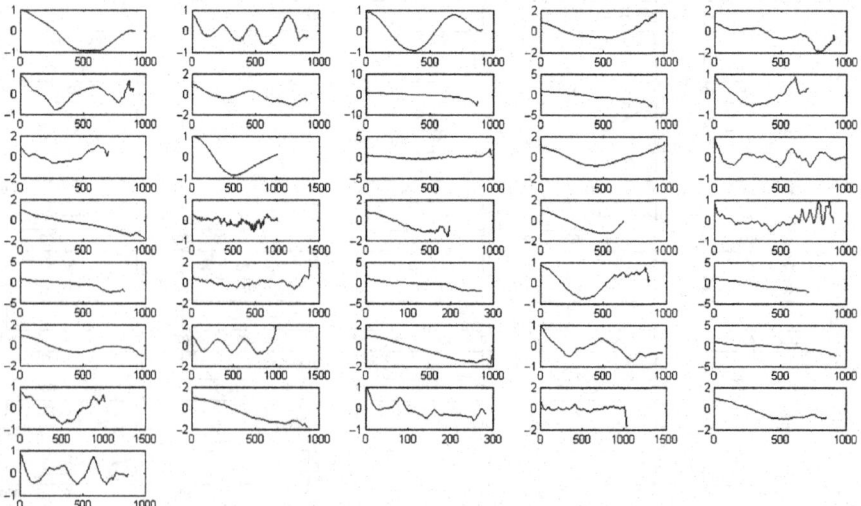

Fig. 8.17 The autocorrelation graphs of the Be stars (Part III) in MACHO. The horizontal axis is the date of observation, while the vertical axis is the computed correlation values of the light curves

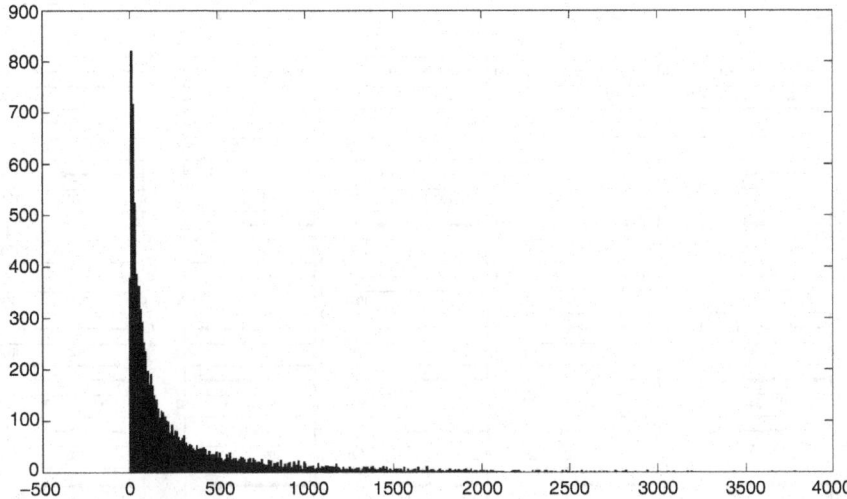

Fig. 8.18 The histogram of the distance between the each pair of the autocorrelations of the Be stars in MACHO. The horizontal axis is the distance values and the vertical axis is the number of observations

Fig. 8.19 The dendrogram of the hierarchical clustering of the autocorrelation sequences of Be stars in MACHO. The labels in the horizontal axis are the assigned Be star number, while the vertical axis is the DTW distance between the Be stars

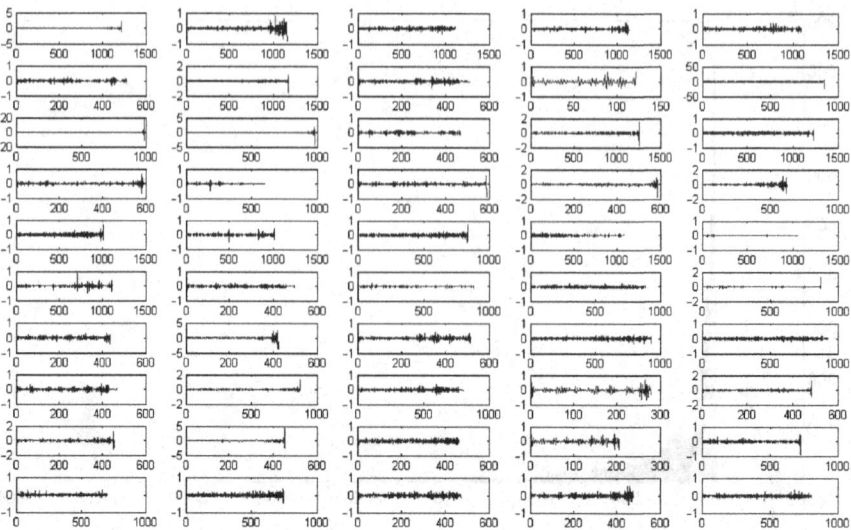

Fig. 8.20 The autocorrelation graphs of the Be stars (Part I) in MACHO (with the high frequency components only). The horizontal axis is the date of observation, while the vertical axis is the computed correlation values of the light curves

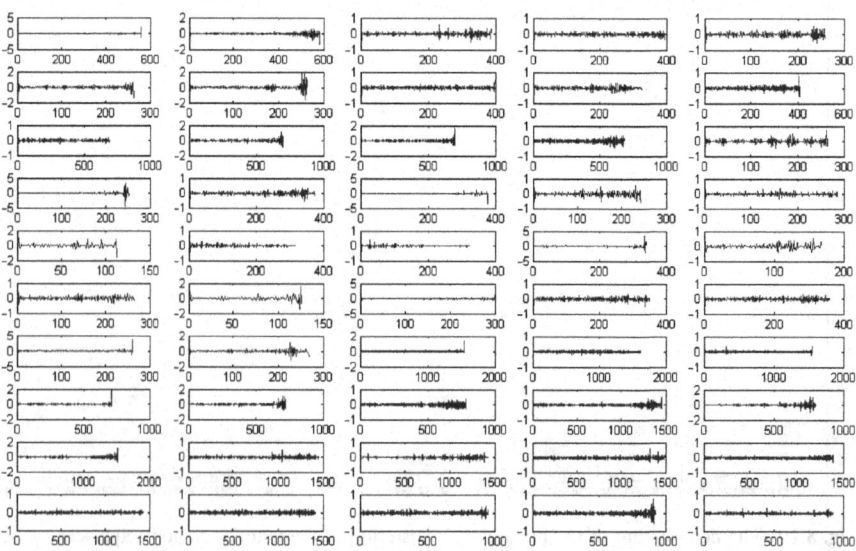

Fig. 8.21 The autocorrelation graphs of the Be stars (Part II) in MACHO (with the high frequency components only). The horizontal axis is the date of observation, while the vertical axis is the computed correlation values of the light curves

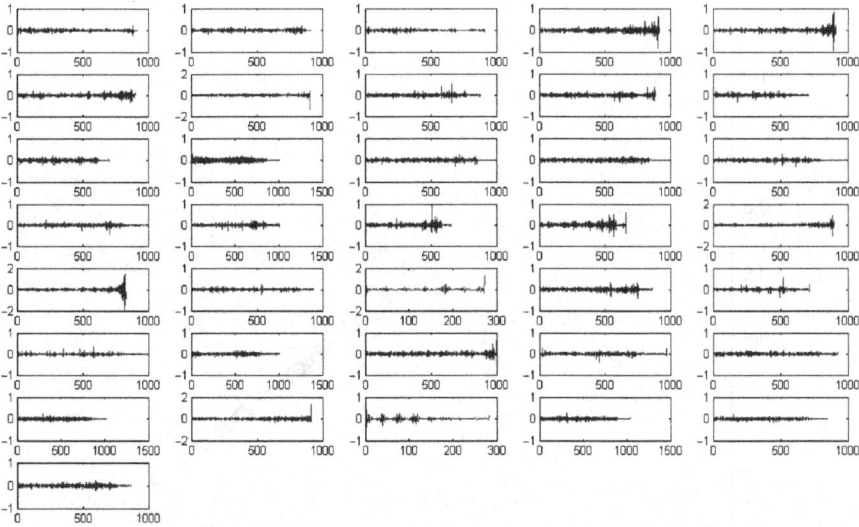

Fig. 8.22 The autocorrelation graphs of the Be stars (Part III) in MACHO (with the high frequency components only). The horizontal axis is the date of observation, while the vertical axis is the computed correlation values of the light curves

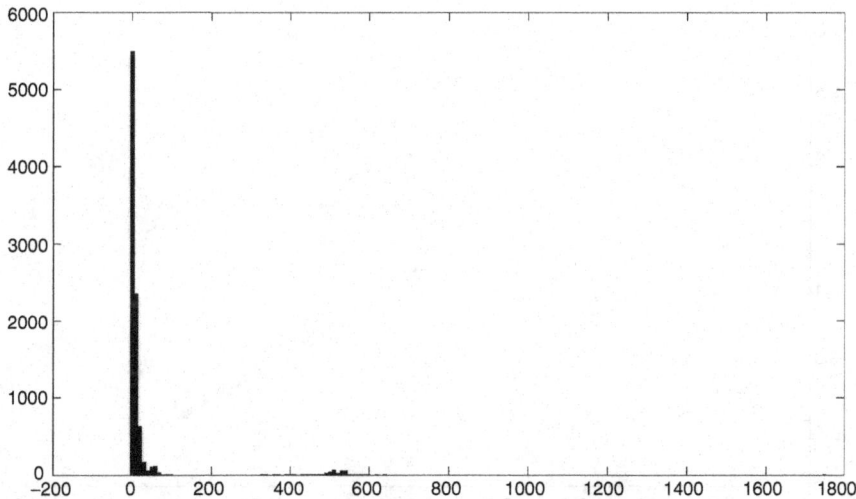

Fig. 8.23 The histogram of the distance between the each pair of the autocorrelations of the Be stars in MACHO (the high-frequency components only)

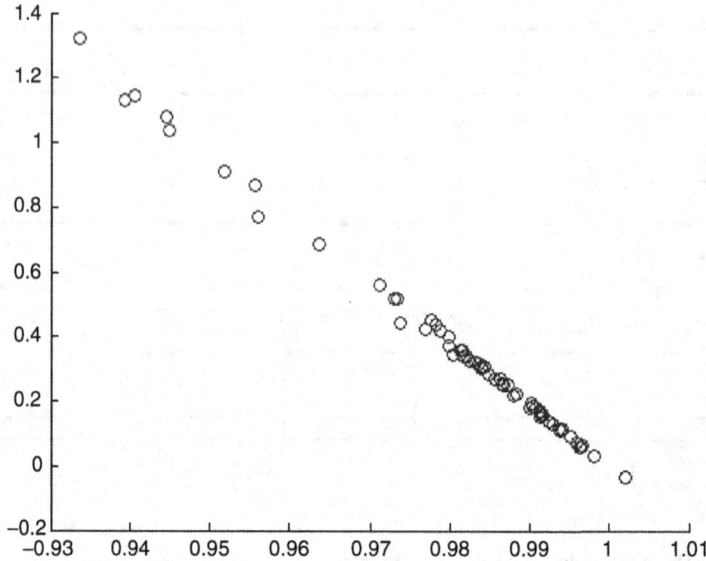

Fig. 8.24 The plots of the coefficients of the constant terms vs. those of the first lag terms of the autoregression equations of the quasars in MACHO. The values constant terms are represented in the horizontal axis and the values of the first lag terms are represented in the vertical axis

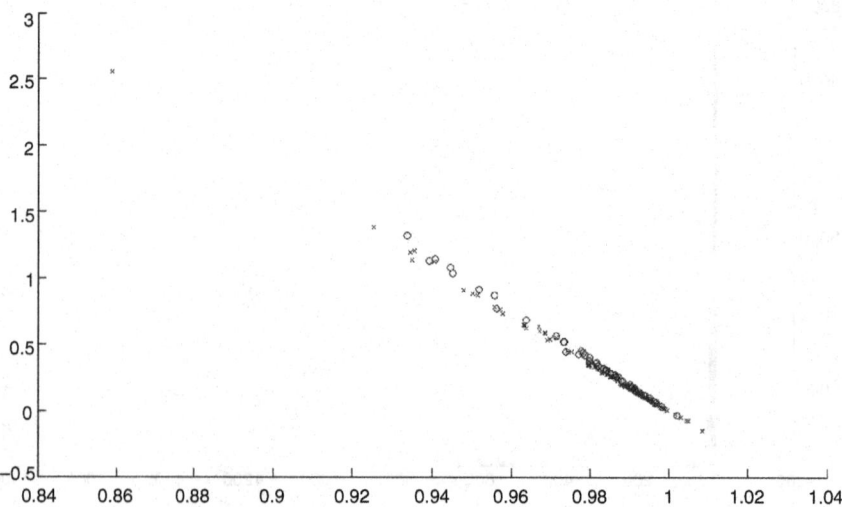

Fig. 8.25 The plots of the coefficients of the constant terms vs. those of the first lag terms of the autoregression equations of the Be stars in MACHO. The values constant terms are represented in the horizontal axis and the values of the first lag terms are represented in the vertical axis

Fig. 8.26 The dendrogram of the hierarchical clustering of the autocorrelation sequences of the quasars and Be stars in MACHO. The labels in the horizontal axis are the assigned quasar/Be star number, while the vertical axis is the DTW distance between the quasar/Be stars

Acknowledgement The author acknowledges that he has talked with his colleague Dr. Pavlos Protopapas, Harvard University, about the properties of the quasars.

References

1. Abazajian et al., 2005, Cosmology and the halo occupation distribution from small-scale galaxy clustering in the sloan digital sky survey. ApJ, 625:613–620.
2. Alcock, C. et al., 2000, ApJ, 542, 281.
3. Bachev, R., 2008, Quasar optical variability: searching for interband time delays, accepted in A&A 2008.
4. Becker, R.H., White, R. L., and Helfand, D. J., 1995, ApJ, 450, 559.
5. Berk, D. et al., 2004, The ensemble photometric variability of ∼ 25,000 quasars in the Sloan Digital Sky Survey, Astrophys. J., 601, 692.
6. Blanco, V. M. and Heathcote, S., 1986, PASP, 98, 635.
7. Croom, S.M., Smith, R. J., Boyle, B. J., Shanks, T., Loaring, N. S., Miller, L., and Lewis, I. J., 2001, MNRAS, 322, L29.
8. Croom, S. M., Schade, D., Boyle, B. J., Shanks, T., Miller, L., and Smith, R. J., 2004, ApJ, 606, 126.
9. Dobrzycki, A., Eyer, L., Stanek, K., and Macri, L., 2005, Discovery of nine quasars behind the large magellanic cloud, A&A, 442, 495.
10. Dobrzycki, A., Groot, P. J., Macri, L. M., and Stanek, K. Z., 2002, ApJ, 569, L15.
11. Dobrzycki, A., Macri, L., Stanek, K., and Groot, P., 2003, Variability-selected quasars behind the small magellanic cloud, AJ, 125, 1330.
12. Eyer, L., 2002, Search for QSO candidates in OGLE-II data, Acta Astronomica, 52, 241.
13. Fan, X., Carilli, C. L., and Keating, B., 2006, ARA&A, 44, 415.
14. Gaskell, C. and Sparke, L., 1986, Astrophys. J., 305, 175.

15. Geha et al., 2002, AJ, 124, 3073.
16. Geha, M. et al., 2003, Variability-selected Quasars in MACHO project magellanic cloud fields. AJ, 125, 1–12.
17. Giveon, U., Maoz, D., Kaspi, S., Netzer, H., and Smith, P. S., 1999, MNRAS, 306, 637.
18. Gray, A. G. and Moore, A. W., 2003, Nonparametric density estimation: Toward computational tractability. In SIAM International conference on Data Mining.
19. Meusinger, H., Scholz, R.-D., Irwin, M., and Lehmann, H., 2002, QSOs from the variability and proper motion survey in the M3 field, A&A, 392, 851.
20. Newberg, H. and Yanny, B., 1997, ApJS, 113, 89.
21. Ratanamahatana, C. and Keogh, E., 2005, Three myths about dynamic time warping data mining. In Proceedings of the 5th SIAM International Conference on Data Mining, Newport Beach, CA, pp. 506–510.
22. Richards, G. et al., 2002, Spectroscopic target selection in the Sloan digital sky survey: The quasar sample, AJ, 123, 2945.
23. Richards, G. et al., 2004, Efficient photometric selection of quasars from the Sloan digital sky survey: 100,000 z < 3 quasars from Data Release One, ApJS, 155, 257.
24. Richards, G. et al., 2009, Efficient photometric selection of quasars from the Sloan digital sky survey: II. ~1,000,000 quasars from Data Release Six, ApJS, 180, 67.
25. Sandage, A. and Wyndham, J. D., 1965, ApJ, 141, 328.
26. Schmidtke, P. C., Cowley, A., Crane, J., Taylor, V., McGrath, T., Hutchings, J., and Crampton, D., 1999, AJ, 117, 927.
27. Schneider, D. P., et al., 2003, The Sloan digital sky survey quasar catalog. II. First data release. AJ, Volume 126, Issue 6, pp. 2579–2593.
28. Schneider, D. P., et al., 2007, AJ, 134, 102, arXiv:0704.0806.
29. Scholz, R., Meusinger, H., and Irwin, M., 1997, A UBV/variability/proper motion QSO survey from Schmidt plates I. Method and success rate, A&A, 325, 457.
30. Schwarzenberg-Czerny, A., 1989, MNRAS, 241, 153.
31. Sumi, T. et al., 2005, Variability-selected QSO candidates in OGLE-II Galactic bulge fields, MNRAS, 357, 331.
32. Theodoridis, S. and Koutroumbas K., 2003, Pattern recognition (2nd edn.). Academic Press, San Diego, CA.
33. Udalski, A., Szymański, M., Kubiak, M., Pietrzyński, G., Soszyński, I., Woźniak, P., Zebruń, K., Szewczyk, O., and Wyrzykowski., 2002, Acta Astron., 52, 217.
34. Vries, W. et al., 2005, Structure function analysis of long-term quasar variability, AJ, 129, 615.
35. Welch and Stetson, 1993, AJ, 105, 1813.
36. Woźniak, P. R., Udalski, A., Szymański, M., Kubiak, M., Pietrzyński, G., Soszyński, I., and Zebruń, K., 2002, Acta Astronomica, 52, 129.

Chapter 9
Improving Low-Cost Sail Simulator Results by Artificial Neural Networks Models

V. Díaz Casás, P. Porca Belío, F. López Peña, and R.J. Duro

Abstract In the present study a method is proposed to reduce the error level of these simplified simulators by correcting the results achieved by means of neural network based approximations. The results of simple aerodynamic simulators used within an evolutionary sail design process are used as application example. The neural network correction is carried out in this case by comparing the numerical results with wind tunnel experiments performed on sail models.

Keywords Aerodynamics · ANN · Wind tunnel · Sail

9.1 Introduction

Fluid dynamic simulations demand complex numerical models and huge computational resources. In some cases this may be regarded as a severe drawback and the search for computational cost reductions may lead to having to rely on simplified models. This situation arises not only in cases where high accuracy is not a priority but also during high level design processes where a large number of proposals must be tested. In any case it is clear that when a simplified model is used the error of the results achieved increases.

The main purpose of any engineering optimization process is to achieve an optimum value for a set of some derived quantities measuring the fitness criteria representing the objective sought in that particular design. The inherent complexity of this process has created a demand for the capability of automating this task; in this sense some proposals making use of different optimization procedures have been published [1, 2]. Evolutionary techniques [3, 4] are the basis of most of these procedures mainly due to their suitability for exploring very complex solution spaces, even in cases where well conformed fitness functions do not exist. The main idea

V.D. Casás (✉), P.P. Belío, F.L. Peña, and R.J. Duro
Integrated Group for Engineering Research, University of Corunna, E.P.S, Ferrol, 15403 Spain
e-mail: vdiaz@udc.es; pporca@udc.es; flop@udc.es; richard@udc.es

S.-I. Ao et al. (eds.), *Advances in Machine Learning and Data Analysis*,
Lecture Notes in Electrical Engineering 48, DOI 10.1007/978-90-481-3177-8_9,
© Springer Science+Business Media B.V. 2010

of automating a design process is to take the designer out of the search/decide loop to avoid the biases it introduces due to previous experience/inexperience and the limited number of alternatives a human is capable of considering.

In most engineering design or optimization problems the objective function or fitness values are often computed using sophisticated and realistic numerical simulations of physical phenomena. Their use within automatic design environments implies the evaluation of a very large number of models – individuals within the evolution process – in order to search design parameter spaces that are characterized by a very high dimensionality. This is an extremely computationally intensive process that may become excessive even when large computational resources are accessible. Consequently, to make them affordable for the computational resources and the available time to be employed in their solution, it is essential to reduce both the number of evaluations and the computational cost of each evaluation as much as possible. The former may be achieved by choosing an appropriate evolution strategy and the latter by using the simplest simulators possible.

In the present research the computational simulations are to be used within an environment for the automatic optimization of sails by means of evolution. As in any case where aerodynamic optimization takes place, the evaluation or calculation of the fitness for each individual or alternative is determined by means of fluid dynamic simulations and therefore it usually involves large quantities of computational resources. Thus the use of simplified and fast simulations is crucial in order to reduce the duration of each simulation and eventually that of the complete optimization procedure. The evaluation of the reliability of simplified CFD codes and the development of appropriate code calibration techniques is of paramount importance in order to be able to decide how faithful to reality their results are. From the point of view of aerodynamic or hydrodynamic design, this is a process that tries to optimize the shape of a body; therefore the model is represented as a surface acting as a boundary on the fluid dynamic problem. Several elements are involved in this evaluation process and all of them affect the final result of the optimization procedure.

The main requirement for the CFD simulators used within the evolutionary process is that of being able to obtain a correct ranking of the sails under evaluation in terms of their fitness. The precision in evaluating the fitness of each individual is unimportant at this stage of the process, because the goal is just to know which individual performs best, that is, to establish a rank of all the individuals in the generation under analysis. Thus, even if the error of the evaluation is large it will not produce any effect on the search process if it does not affect de achievement of this main ranking requirement. It is important to point out here that the literature provides information about the accuracy of different CFD codes [5, 6], but not on their ability to perform good fitness ranking. It is in this framework where the appropriate experimental validation procedure becomes so important in order to determine the quality of the results.

In this work we are going to present some simplified models for the evaluation of sails within evolutionary design procedures [7, 8] and a wind tunnel based methodology for the validation of the results provided by this model. Special emphasis will be placed on the description of neural network procedures for improving the results of the simplified fluid dynamic models by considering the experimental data used for validation. A general description of the evolutionary environment within which the simulator and the validation strategy operate may be found in Ref. [7].

9.2 Fluid Dynamics Simulation

As stated earlier, in order to reduce the computational cost of any aerodynamic optimization problem, such as the one we are dealing with, either the number of evaluations must be reduced, or the duration of each one shortened. The first possibility may be achieved in two different ways:

- Decreasing the dimensionality of the problem, i.e. the degrees of freedom of each alternative
- Reducing the size of each dimension through a smaller resolution or a reduced range

However, both ways imply disregarding some of the possible solutions. Consequently, to allow for an unconstrained search domain, we have concentrated on the second possibility, that is, on the reduction of the computational cost of the simulator. The repercussion of this approach on the quality of the results of the aerodynamic simulator needs to be evaluated and it is mainly influenced by two factors:

- The fluid dynamics numerical model
- The mesh size

The mesh element size determines the equation system size in the discretized scheme. This size is proportional to the number of mesh nodes, which for a given domain grows inversely to the cube of mesh element size.

The fluid dynamics are governed by the Navier-Stokes equations. However, the complete system generated by the direct application of the Navier-Stokes equations cannot be reasonably solved in complex real problems. Thus the fluid dynamics numerical model used must be adapted to the characteristics of the problem. The first step in this process is to study the mean parameters defining the fluid motion around the sail [9].

In the particular case of sails, which is the problem considered here, the characteristic dimensions are:

- Kinematic air viscosity at 25°C (ν):1.5 10^{-5} m^2/s
- Wind velocity (v): 10 m/s
- Sail characteristic length (L): 12 m

From these, the values of the two standardized non dimensional parameters characterizing the problem can be obtained. These are the Reynolds number, with a value of $1.3 \cdot 10^7$ and a Mach number of 0.029.

Additionally, the fluid can be considered incompressible if

$$M \ll 1, \quad \frac{M^2}{Fr^2} \ll 1, \quad \frac{M^2}{Re} \ll 1. \tag{9.1}$$

These conditions are fulfilled in this case and therefore air can be considered as incompressible in these simulations.

Another simplification that may be made is to neglect viscous forces. Simple fluid dynamic considerations indicate that if the Reynolds number is significantly larger than one, this simplification may be made. In such a case the Navier-Stokes equation system is reduced to the Euler stationary equation by assuming an incompressible and inviscid flow:

$$\vec{v} \cdot \nabla \vec{v} + \frac{\nabla p}{\rho} = 0 \tag{9.2}$$

Therefore, using a larger mesh and solving the Euler equation instead of the Navier-Stokes equation the computational cost may be drastically reduced. Obviously, the reduction is achieved at the expense of solution accuracy.

This is reflected by the fact that:

- The use of larger mesh element size increases de size of the smallest fluid structure that may be captured by the simulator.
- The fluid can be taken as incompressible and inviscid in large regions of the domain. However in certain areas, such as in the proximity of the sail and in turbulent wakes, these considerations are not valid. Thus, in regions with important changes in the value of the velocity gradient, where the dissipation processes appear, the value of the Reynolds number and the value of the Laplacian of the velocity are of similar order and the viscous effects must be taken into account.

These considerations have been introduced in the aerodynamic simulator developed for this work in order to reduce the computational cost of each simulation. However, errors are introduced in the results and as a consequence the solutions must be validated through experimental procedures in order to insure that they reflect reality.

9.3 Sail Shape Definition

When trying to automate a design process, as is the case here for sails, the main idea is that any design process can be restated as an optimization procedure. Thus, when optimizing, one is mainly searching through a high dimensional space corresponding to the parametric definition of that to be optimized, testing or evaluating different points or alternative designs within this solution space and trying to combine the information obtained from these tests to seek points or alternative designs that are potentially better. Consequently, there are two main processes involved. On one hand, we need a search procedure that allows us to move throughout the high dimensional solution space and make the best possible use of the information provided by each test or sample and, on the other, we must be able to evaluate each sample somehow, usually through a simulator.

In the case of sail design, each stage of the optimization process uses a different representation of the alternative: the search stage, which is here an evolutionary process, uses chromosomes obtained from a parametric definition of the sail, and the simulator that provides the evaluation of each alternative uses the three-dimensional representation of the sail surface. This makes it necessary to introduce a decoding stage enabling communication between different modules that make up the design environment. There is a huge number of different ways to achieve a good approximation to the definition of the surface. However, the parameters used to define the surface will be used in the optimization process in order to find the best sail shape. Consequently, the criteria employed to choose the numerical model of the surface are:

- **Low number of parameters**: Using a small number of parameters reduces the size of the search space.
- **Simple definition of the surface**: Allowing for an easy interpretation of the results in the first steps of the design process and resulting in a lower computational cost than a complex definition.

As stated earlier, to facilitate the definition and control surfaces, non uniform b-splines (NURBS) have been considered. This way the shape of the sail surface is given by a set of control points, except those at the border which are not considered as a part of the surface. These control points act as attractors of the surface, which allows for an easy way to control the shape and concavity by changing their position and weight.

Consequently, the number of parameters used to define the shape can be easily adapted to the requirements of each experiment. These parameters define the position and the weight of each control point, directly or through a behavior equation. As commented earlier, this procedure makes it necessary to add a process of integration of the values of length and area to ascertain the real values over the sail.

The sails selected to perform these experiments are of the Olympic "Tornado class" and the international rules for this class have been used to define constraints for the sail optimization process. These constraints determine the maximum size and the relative position between the main sail and the jib.

The limits on the size are defined over the maximum length of the sail at different heights. Thus, before the aerodynamic analysis, these values are calculated and if all the constraints are not met, the sail is rejected.

To define the main sail shape, five parameters have been used:

- Maximum displacement and its position
- Maximum displacement in the head
- Displacement of the trailing edge in the head
- Displacement leech over the foot
- Line of maximum deformation over the sail

To define the jib, three parameters have been used:

- Maximum displacement and its position
- Displacement of the trailing edge
- Line of the maximum deformation

In addition, a parabolic definition for the weight and for the horizontal position of the control points was used in order to increase the range of alternatives that can be evaluated.

9.4 Experimental Validation

The aim of the evaluation process in an evolutionary design strategy is to determine a correct rank among the different alternatives under evaluation. Thus, the results of the simulator must be qualitatively acceptable even though they may not be quantitatively realistic. This is a considerable advantage but one must be aware that it is quite doubtful that a highly inaccurate simulator may be able to produce adequate ranks.

In this study, the quality or fitness of each individual is determined by the values of the components of the total force over the sail. The experimental validation of the results of the simulator implies comparing the real forces and the forces calculated by the numerical simulator. Basically, it consists in the evaluation of a sail configuration under different angles of attack and comparing the evolution between real and numerical results. To compare these two results it is necessary to determine both the real shape of the sail and the forces acting over it.

9.4.1 Experimental Apparatus

Tests were performed on a sail model placed in the subsonic closed loop wind tunnel of the University of Corunna Fluid Dynamics Lab. This wind tunnel has an open test section of 1.1×0.8 m, its wind velocity can be continuously controlled and adjusted from 5 to 45 m/s, the turbulence level is below 1% and the maximum value in axial velocity non uniformity is of 2%. The sail is supported by a six component force and moment transducer. This set-up is complemented by a smoke tracer for flow visualization and a particle image velocimeter (PIV).

9.4.2 Sail Shape Analysis

The first step in the simulator validation process is to determine the real shape of the sail. The sail is an anisotropic membrane and solving the problem of the real sail shape under a specific wind condition implies solving a highly non-linear problem of fluid-structure interaction. To eliminate this problem the sail is considered as a solid body in the simulator.

The shape of the sail must be measured under each specific wind condition in order to be introduced in the simulator. The measurement process must be non-intrusive and external to the sail to avoid disturbing the flow around the sail. In this case a stereoscopic camera system has been proposed to acquire the sail shape by means of photogrammetric techniques. This process can be divided into the following stages:

- **Sail marking**. A grid of point is marked on the sail in order to permit an easy recognition of its surface.
- **Photography**. Simultaneous photographs are taken from different points of view. In the case presented in Fig. 9.1 three different photographs were used. The use of simultaneous pictures eliminates the possible distortions introduced by sail movements during the process.
- **Point recognition**. The marked grid is recognized in each photograph and the three-dimensional grid is regenerated by obtaining the correspondence between points in all of the images.
- **Surface definition**. A surface is generated through a minimum least square method using the points of the grid and applying all the constraints of the problem. These constraints are the fixed position of the mast and the fact that the sign of the convexity over the whole surface must be preserved.

9.4.3   Forces Analysis

The sail position is defined by the mast and boom position. The mast is supported by a platform that permits the simulation of the relative course to the wind and the roll of the ship. The connection between the mast and the platform is achieved through the six axes force/torque transducer, which provides the forces and moments on

Fig. 9.1 Photographs taken by each camera in the wind tunnel

each axis. The boom is linked to the mast by a spherical joint and its position is determined by the length of the sheet. The control of its length is carried out by a winch over a strain-gage cell that measures the tension (T) in the sheet. Through this distance, the roll of the platform and the measurement of the direction of the sheet are calculated, a director vector is obtained as:

$$\vec{s} = (\cos(\alpha), \cos(\beta), \cos(\gamma)), \tag{9.3}$$

where α is the angle between the sheet and the X axis, β is the angle between the sheet and the Y axis and γ is the angle between the sheet and the Z axis. The procedure used to calculate these angles has been designed as a redundant system in order to reduce the error in their values.

Using the mast and boom position and the sheet direction, all forces and moments are obtained. Therefore the center of pressure and the total force may be calculated. Thus, the total force acting over the sail is evaluated as

$$\begin{aligned}
R_x &= F_x + T \cdot \cos(\alpha) \\
R_y &= F_y + T \cdot \cos(\beta). \\
R_z &= F_z + T \cdot \cos(\gamma)
\end{aligned} \tag{9.4}$$

The position of the center of pressure is calculated through the relation between the moment and the total force.

9.5 Neural Network Correction

9.5.1 Integrating Computational and Experimental Analyses

Given the wide range of angles, shapes and relative positions analyzed during the optimization process imply that several considerations assumed in the numerical model may not be applicable. When this happens, the accuracy of the simulator is compromised, especially if a large high turbulence region is present. To correct the deviations from the real values of lift and drag forces produced by the sail, a neural network is used within our scheme. We made use of a simple multilayer perceptron neural network [10] with 7 input neurons, two hidden layers having 12 neurons each and 2 output neurons (lift and drag coefficients). The input neurons correspond to: calculated lift coefficient, calculated drag coefficient, air velocity, angle of attack, and eight values defining the shape of the sail (weight and position of control points). This set of inputs contemplates two classes of parameters: two that depend only on the case being studied (attack angle and Reynolds number) and four that depend on the sail profile.

The validation process is carried out by means of a systematic series of measurements taken for several courses and roll angles. For each combination of course and roll the sail shape under the action of the wind is determined by a photogrammetric procedure as explained above. It is then numerically simulated with the same shape and under the same conditions. The values of the forces and moments measured by the six axis balance supporting the mast and the strain-gage cell value for the tension in the boom rope are used as fitness parameters to evaluate the performance of the simulation.

To define the ANN correction, test results were divided into two groups: a training set (four samples) and a test set (five samples). The training was carried out through a gradient descent method with relaxation parameters of 0.3. In the proposed case, the training ended with an iteration limit of 2,000 iterations, or when the relative error was less than 0.05%.

Figures 9.2, 9.3 and 9.4 show the results of the ANN correction. The maximum error in the test cases is 1.2% and the mean error is 0.23%. The use of the ANN correction reduced the mean error from 11.54% to 0.23%. This correction allows realistic results to be obtained without including the hull and the mast in the simulator, and using a low computational cost solver.

Although the simulation stage demands large computational resources, due to the complexity of the aerodynamic behaviour of the sails, ANN correction requires its

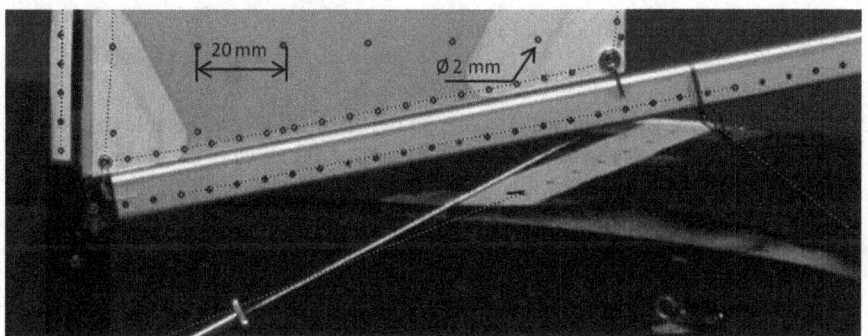

Fig. 9.2 Characteristic dimensions of the reference marks on the mast and the foot

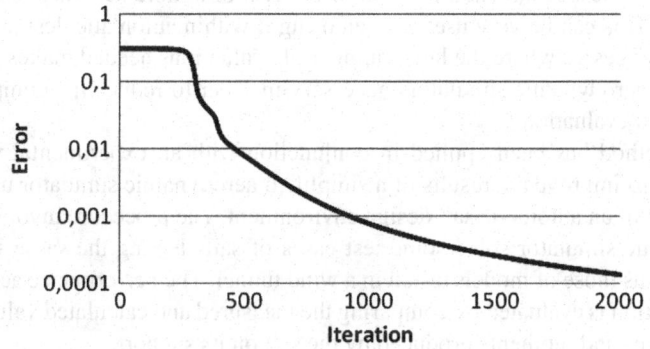

Fig. 9.3 Evolution of the error of the ANN during the training process

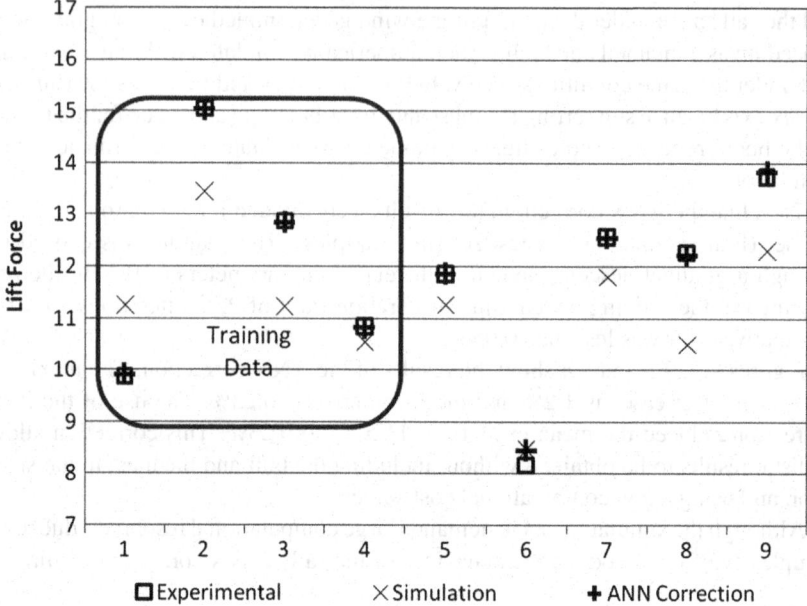

Fig. 9.4 Results of the ANN for the lift and drag force

results as reference values in order to correct the error values and be able to consider the effects commented above. That is if only shape parameters are introduced, the ANN's cannot obtain the force functions directly.

9.6 Conclusion

A neural network based method for improving the results obtained with simplified aerodynamic simulators has been developed. The results show that this method can improve these results at a fraction of the cost of a more accurate aerodynamic simulator. This can be very useful when using it within automatic design and optimization processes where the large number of evaluations needed makes the use of simplified aerodynamic simulators necessary in order to reduce the computational cost of each evaluation.

The method has been applied in conjunction with an experimental validation procedure to improve the results of a simplified aerodynamic simulator used in the framework of an automatic sail design environment. The procedure involves having aerodynamic simulator solve some test cases of sails having the same shape and conditions as those of models tested in a wind tunnel. The performance achieved by the simulation is evaluated by comparing the measured and calculated values for the global forces and moments produced by the sail on its support.

References

1. Ziman J. (2003) Technological Innovation as an Evolutionary Process, Cambridge University Press, Cambridge.
2. Nolfi S., Floreano D. (2004) Evolutionary Robotics: The Biology, Intelligence, and Technology of Self-Organizing Machines, Bradford, Cambridge, MA.
3. Michalewicz, Z., Dasgupta, D., Le Riche, R.G., Schoenauer, R. (1996) Evolutionary algorithms for constrained engineering problems, Computers & Industrial Engineering, Vol. 30, No. 4, pp. 851–870.
4. Gen, M., Cheng, R. (1998) Genetic Algorithms and Engineering Design, Wiley, New York.
5. Mehta, U.B. (1998) Credible computational fluids dynamics simulations, AIAA Journal, Vol. 36, pp. 665–667.
6. Stern, F., Wilson, R. V., Coleman, H., Paterson, E. (2001) Verification and validation of CFD simulations: Part 1 – comprehensive methodology, ASME Journal of Fluids Engineering, Vol. 123, pp. 793–802 G.
7. Lopez-Peña, F., Diaz Casas, V., Duro, R. J. (2008) Automatic Aerodynamic and Hydrodynamic Design Based on Simulation and Evolution, Tenth International Conference on Computer Modeling and Simulation, UKSIM 2008, Cambridge, UK, pp. 7–13, 1–3 April 2008.
8. Díaz Casás, V., Duro, R. J., Lopez Pena, F. (December 2005) Evolutionary design of wind turbine blades, International Scientific Journal of Computing, Vol. 4, No. 3, pp. 49–55.
9. Jerome H. Milgram (1998) Fluid mechanics for sailing vessel design, Annual Review of Fluid Mechanics, Vol. 30, pp. 613–653.
10. Haykin, S. (1999) Neural Networks: A Comprehensive Foundation, Prentice Hall, Upper Saddle River, NJ.

Chapter 10
Rough Set Approaches to Unsupervised Neural Network Based Pattern Classifier

Ashwin Kothari and Avinash Keskar

Abstract Unsupervised neural network based pattern classification is a widely popular choice for many real time applications. Such applications always face challenges of processing data with lot of consistency, inconsistency, ambiguity or incompleteness. Hence to deal with such challenges a strong approximation tool is always needed. Rough set is one such tool and various approaches based on Rough set, if are applied to pure neural (unsupervised) pattern classifier can yield desired results like faster convergence, feature space reduction and improved classification accuracy. The application of such approaches at respective level of implementation of neural network based pattern classifier for two case studies are discussed here. Whereas more emphasis is given on the preprocessing level based approach used for feature space reduction.

Keywords Discernibility · Feature extraction · Pattern classification · Reducts · Rough neuron · Rough sets · Unsupervised neural network

10.1 Introduction

Rough sets theory exploits the inconsistency and hidden patterns present in the data. Rough sets have been proposed for a variety of applications. In particular, the rough set theory approach seems to be important for artificial intelligence and cognitive sciences, especially for machine learning, knowledge discovery, data mining, expert systems, approximate reasoning and pattern recognition. Artificial Neural Networks in the most general form aim to develop systems that functions similar to the human brain. The nature of connections and the data exchange in the

A. Kothari (✉)
Senior Lecturer, Department of Electronics & Computer Science, VNIT, Nagpur, India
e-mail: agkothari72@rediffmail.com

A. Keskar
Professor and Dean R&D, Department of Electronics & Computer Science, VNIT, Nagpur, India

S.-I. Ao et al. (eds.), *Advances in Machine Learning and Data Analysis*,
Lecture Notes in Electrical Engineering 48, DOI 10.1007/978-90-481-3177-8_10,
© Springer Science+Business Media B.V. 2010

network depend on the type of application. Rough sets and Neural Networks can be combined, as they would be effective in cases of real world data, which are ambiguous, imprecise, incomplete and error prone.

Here two rough-neuro based hybrid approaches are proposed for classification by unsupervised ANN. As the rough set theory can be used for reducing the input feature space for the neural network doing the classification, first approach is suggested at preprocessing level. The second approach respectively explores the application of rough set theory for architectural modifications in the neural network. The first case study of printed character recognition has been undertaken to establish that a Rough-Neuro Hybrid approach has reduced dimensionality and hence also has resulted in lesser computations as compared to the pure neural approach. The data set used consists of characters A–Z, in 18 different fonts. Whereas second case study is focused on application of above discussed approaches for hand off prediction for cellular communications. The results for pattern classification using a Pure Neural approach have been used for benchmarking for both cases. Hence nature of information system (IS) and futures used in both cases along with the steps of image preprocessing and Feature extraction are discussed in initial sections. As the values obtained are continuous, the discretization steps used are explained in the following section. The feature space reduction and rough hybrid approach are discussed in the subsequent sections. The last section presents the results and conclusions.

10.1.1 Basics of Rough Set Theory

In this section some of the terminologies related to the basics of rough set theory and frequently used in the subsequent sections are explained [1].

10.1.1.1 Information System (IS)

The basic vehicle for data representation in the rough set framework is an information system. An information system is in this context a single flat table, either physically or logically in form of a view across several underlying tables. We can thus define an information system I in terms of a pair (U, A), where U is a non-empty finite set of objects and A is a non-empty finite set of attributes. The input feature labels are termed as condition attributes whereas class label is termed as decision attribute. The Table 10.1 shown below is the sample of IS used for second case study. The seven columns or input features namely BST2, BST3, BST4 (base station id), RSS (received signal strength), TMSTMP (time stamp), SPEED (speed of user motion) and TIME (time) are the condition attributes while the output class column CLASS is the decision attribute.

Table 10.1 Sample information system (IS)

BST2	BST3	BST4	RSS	TMSTMP	SPEED	TIME	CLASS
5	6	5	80	20	50	10	1
1	1	6	75	36	28	10	3
4	3	2	72	26	23	10	5
4	4	4	72	18	49	10	4
5	5	3	81	34	32	10	2

10.1.1.2   Reducts

The reduced sets of attributes are called as Reducts. Reducts derived out of input vectors using rough set postulations ease the process of making predictions and decision making which in turn gives improved classification with reduced dimensionality of feature space. Whereas theoretically reducts are defined as follows,

A subset B of set A is a reduct if and only if,

- $B^* = A^*$.
- B with this property i.e. $(B - \{a\})^* \neq A^*$ for all $a \in B$, $*$ is the partition on U because of indiscernibility.

Once reducts are known, rules can be easily generated for classification.

10.1.1.3   Core

Cores are the prime attributes or Indispensable condition attributes. Core is the set of all those attributes, which are essential for classification between two classes, and there is no alternative for those attributes. If core is not included in the reducts then efficiency dramatically decreases. For example in the above shown sample IS, BST2, BST3 and BST4 are some of the core attributes.

10.1.1.4   Discernibility Matrix

It is an information system I defines a matrix MA called a discernibility matrix. Each entry MA (x, y) which is subset of A consists of the set of attributes that can be used to discern between objects x, y which are elements of U.

10.2   Image Processing and Feature Extraction for the First Case Study

The original data set is subjected to a number of preliminary processing steps to make it usable by the feature extraction algorithm. Pre-processing aims at producing data that is easy for the pattern recognition system to operate accurately. The main objectives of pre-processing [2] are: binarization, noise reduction, skeletonization, boundary extraction, stroke width compensation [3], truncation of redundant portion of image and resizing to a specific size. Image binarization consists of conversion of a gray scale image into a binary image. Noise reduction is performed using morphological operations like dilation, erosion etc. Skeletonization of the image gives an approximate single-pixel skeleton, which helps in the further stages of feature extraction and classification. The outermost boundary of the image is extracted to further obtain the boundary related attributes such as chain codes and number of loops. Stroke width compensation is performed to repair the character strokes, to fill small holes and to reduce uneven nature of the characters. The white portion surrounding the image can create noise in the feature extraction process and also increases the size of image unnecessarily. Truncation is performed to remove this white portion. In the end, the image is resized to a pre-defined size: 64 × 64 pixel in this case. All such results are indicated in Fig. 10.1 below. In feature extraction stage, each character is represented as a feature vector, which becomes its identity. The major goal of feature extraction is to extract a set of features, which maximizes the recognition rate with the least amount of elements. The feature extraction process used consists of two types of features: statistical and structural [3–5]. The major statistical features used are: zoning, crossings, pixel density, Euler number, compactness, mean and variance. In zoning, the 64 × 64 character image is divided into 16 × 16 pixel parts and pixel density of each part is calculated individually. This helps in obtaining local characteristics rather than global characteristics and is an important attribute for pattern recognition. Crossings count the number of transitions from background to foreground pixels along vertical and horizontal lines through the character image. In Fig. 10.2 there are six vertical crossings (white to black and black to white) and four horizontal crossings in both the upper and lower part. Pixel Density is calculated over the whole 64 × 64 image. Euler number of an image is a scalar whose value is the total number of objects in the image minus the total number of holes in those objects. Euler number is also calculated for each image. Structural features are based on topological and geometrical properties of the character, such as aspect ratio, loops, strokes and their directions etc. The boundary of the image is obtained and chain code is calculated for it. Then the number of ones, twos till number of eights is calculated. The number of loops present in a character is also obtained.

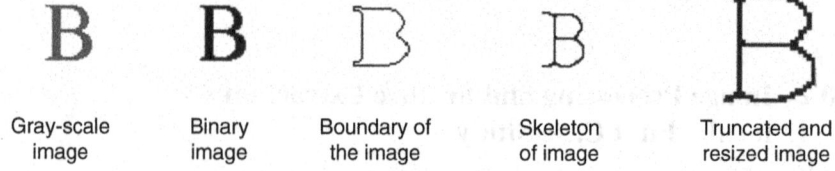

| Gray-scale image | Binary image | Boundary of the image | Skeleton of image | Truncated and resized image |

Fig. 10.1 Various stages of image processing

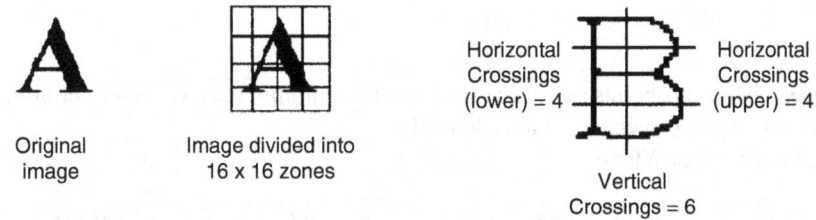

Fig. 10.2 Image segmentation and feature extraction

10.3   Discretization

The solution to deal with continuous valued attributes is to partition numeric variables into a number of intervals and treat each such interval as a category. This process of partitioning continuous variables is usually termed as discretization. Data can be reduced and simplified through discretization. In case of continuous valued attributes, large number of objects are generated with a very few objects mapping into each of these classes. The worst case would be when each value creates equivalence with only one object mapping into it. The discretization in rough set theory has particular characteristic, which should not weaken the indiscernibility ability. A variety of discretization methods have been developed.

10.3.1   Algorithm

Consider an information system $IS = (U, A, V, f)$ where U: is the universal sets containing all objects i.e. $U = \{x_1, x_2, x_3, \ldots, x_n\}$, n is the total number of objects; $A = C \cup \{d\}$ where C denotes the set of condition attribute and d is the decision attribute; V denotes contains the sets of values each condition attribute can take; f is a function between the element in U and its value, the value of object x_i to the attribute a is $a(x_i)$. The process of discretization is to find the sets of cut points for each attribute and hence discretize each of them. For an attribute a, V_a the set containing the values the attribute can take. Cuts are nothing but partitions of the set V_a i.e. $V_a = [c\ d]$, where $[c\ d]$ is the interval of the continuous valued attribute. Then a partition: $c < p_1 < p_2 \ldots \ldots < p_m < d$, where the set $\{p_1, p_2, \ldots . p_m\}$ forms the set of cut points which divides the interval $[c\ d]$ into m intervals without intersection: $[c\ p_1), [p_1\ p_2), [p_2\ p_3) \ldots . (p_m d]$ and the continuous values of attribute a turn out to be $m + 1$ discrete values: $V_1, V_2, V_3, \ldots \ldots, V_{m+1}$ by the following equations [6]:

$$\{V_1, \quad \text{if } a(x) \le p_1\} \tag{10.1a}$$
$$V(x) = \{V_i, \quad \text{if } p_{i-1} < a(x) \le p_i, i = 1, 2, 3 \ldots, m\} \tag{10.1b}$$
$$\{V_m{+}_1, \quad \text{if } a(x) > p_m\} \tag{10.1c}$$

10.3.2 Steps to Obtain Cuts

To search the cuts points we make use of the discernibility matrix. The discernibility matrix of a given decision table is defined as:

$(M_{\{d\}}(i, j))_{nxn}$ Where

$$M_{\{d\}}(i, j) = \{\{a_k | a_k(x_i) \neq a_k(x_j), \text{ for all } a_k \text{ in C}\} \text{ when } d(x_i) \neq d(x_j)\}$$
$$\{0, d(x_i) = d(x_j)\} \hspace{4cm} (10.2)$$

10.3.3 Steps for Algorithm

1. Construct the discernibility matrix for the given table
2. For $i = 1$ to t
3. For all $1 \leq j \leq k \leq n$, if the attribute a_i is a member of the discernibility matrix entry, construct the cuts interval $[a_i(x_j)\ a_i(x_k)]$
4. For every set of intersecting intervals, construct the cut point as:
   $(p_j)_i = \{\max(\text{lower bounds}) + \min(\text{upper bounds})\}/2. (3)$
5. Discretize attribute a_i according to the cuts obtained
6. Next i
7. End

The Figs. 10.8 and 10.9 show an undiscretized table and the discretized table using the above-discussed algorithm. It is observed that the continuous interval valued attributes are discretized without affecting their ability to discern. It can be further seen that this helps in attribute reduction also.

10.4 Reduction of Attributes

Use of rough sets theory in the preprocessing stage results into dimensionality reduction and optimized classification with removal of redundant attributes. Also, neural network is the most generalized tool for pattern recognition and has capability of working in noisy conditions also. Here a new Rough-Neuro Hybrid Approach in the pre-processing stage of pattern recognition is used. In this process, a set of equivalence classes, which are indiscernible using the set of given attributes, are identified. Only those attributes are kept which preserve the indiscernibility relation and the redundant ones are removed, as they do not affect the classification. A reduction is thus resulting in a reduced set of attributes, which classifies the data set with the same efficiency as that of the original attribute set.

10.4.1   Steps for Finding Reduced Set of Attributes

Consider an information system $IS = (U, A, V, f)$ where U: is the universal sets containing all objects i.e. $U = \{x_1, x_2, x_3, \ldots, x_n\}$, n is the total number of objects; $A = C U \{d\}$ where C denotes the set of condition attribute and d is the decision attribute; V denotes contains the sets of values each condition attribute can take; f is a function between the element in U and its value, the value of object x_i to the attribute a is $a(x_i)$. The total number of condition attributes is m i.e. $|C| = m$. The number of decision classes is t i.e. $|V_d| = t$. Discernibilty matrix $(t \times t)$, whose entries contain the relative significance of each attribute. In this method the relative significance of each attribute in discerning between two compared classes x and y is as given by:

$$P(x)_{x,y}|a_i \qquad (10.3)$$

where $P(x)_{x,y}|a_i$ is the probability of an object belonging to class x given the only information as attribute $a_i (i = 1, 2, 3, \ldots, m)$ when discerning the objects of x from y. If the value turns out to be 1, then the particular attribute is the most significant and if the value turns out to be 0, then the particular attribute is the least significant. The probability described is subjective and there can be other viewpoint. The rough sets deal with uncertainty of data sets or information granules.

10.4.2   Algorithm for Finding Reducts

The algorithm used is as follows:
1. Obtain the information system of which the feature space is to be reduced.
2. The discrenibility matrix (t × t) is to be constructed and entry for each attribute is given by the above method.
3. The relative sum for each attribute is obtained i.e. the contribution of each attribute over the table is summed up.
4. The most significant contributors based on the relative sum are selected according to the requirement or set threshold.

10.5   Rough Neuro Based Hybrid Approach and the Experimentation Done

As discussed in the introduction Rough set can deal with consistent, inconsistent and even incomplete type of data. Hence the methodology proposed must be tested for all the above-mentioned types of data. Hence to examine the outcome for consistent type of data case study one is considered in which the approaches are tested for printed character recognition. While for testing outcome for inconsistent or incomplete data case study 2 is considered in which approaches are used for hand off prediction. In the coming era of cellular communications the cell size will reduce and in situations like mobile user using mass rapid transit systems, the overall performance of the system can be improved if the network can predict the next cell to which user is going to move. Because dynamic cellular traffic conditions and numerous path profiles of moving mobile users data generated will be highly inconsistent in nature.

For both the cases the results achieved by the hybrid approach are compared with the pure neural approach. The type of ANN used in both the case studies for pure and rough-neuro hybrid approach is unsupervised with two layers. When pure neural approach is used for the first case study, the input layer consisting of 31 nodes for 31 input features, fed for classification is used. Whereas the output layer contains 26 nodes for 26 output classes in which the input samples are to be classified.

Whereas for second case study the input layer consists of seven nodes for seven input features and the output layer contains six nodes representing six possible destinations for each current cell of the user location for predicting the handoffs. The data table used for training is formulated with the steps explained earlier. Many such tables with variations in data patterns were used for training and testing in 80:20 proportion. The training algorithm used is (of competitive learning type) Winner take all [7–10]. Rough set approach used for preprocessing or attributes reduction and the downsized set of attributes can be fed to the neural classifier as shown below in Fig. 10.5. Reducts derived out of input vectors using rough set postulations ease the process of making predictions and decision making which in turn gives improved classification with reduced dimensionality of feature space. For estimation of reducts discernibility matrix is first calculated, weighted contributions obtained and the significant contributors are selected as per the fixed threshold. Thus rough set mainly exploits discernibility and hidden inconsistency in the data. The results shown here are for the same table used earlier for the first approach in Fig. 10.3 and Fig. 10.4.

Attribute→	1	2	3	4	5	6	7	8	9	10
1	0.056274	4	2	2	2	1	0	69	69	0
2	0.1073	4	3	2	2	2	106	136	88	110
3	0.066528	−1	2	2	2	0	70	103	64	109
4	0.104919	−1	3	3	3	1	201	122	77	0
5	0.120117	1	4	2	2	0	247	112	112	112
6	0.111206	5	3	3	1	0	110	242	44	202
7	0.10791	4	4	2	2	0	19	172	141	87
8	0.136475	8	1	2	2	0	160	132	132	164
9	0.25	8	1	2	2	0	256	256	256	256
10	0.057068	2	1	1	1	0	0	0	0	176
11	0.1297	6	1	2	2	0	160	144	58	29
12	0.097595	5	2	1	2	0	88	232	16	0
13	0.137634	5	1	4	4	0	117	193	32	256
14	0.114502	4	1	3	4	0	236	34	0	160
15	0.09198	4	2	2	3	1	48	152	141	126
16	0.101807	3	3	3	2	1	229	112	120	137
17	0.092346	4	3	2	4	1	63	134	132	91
18	0.108459	5	2	2	3	1	200	128	133	100
19	0.08783	2	3	1	2	0	103	98	102	101
20	0.062134	0	2	1	1	0	112	157	157	112
21	0.077271	0	1	2	2	0	144	0	0	144

Fig. 10.3 Undiscretized data for case study-1

Attribute→	1	2	3	4	5	6	7	8	9	10
1	1	15	3	3	3	3	1	21	29	1
2	70	15	5	3	3	5	30	55	34	42
3	11	5	3	3	3	1	10	34	26	41
4	67	5	5	5	5	3	73	44	32	1
5	81	9	7	3	3	1	85	37	42	45
6	76	17	5	5	1	1	33	81	18	81
7	72	15	7	3	3	1	5	69	66	24
8	94	22	1	3	3	1	58	51	58	71
9	99	22	1	3	3	1	86	82	86	87
10	2	11	1	1	1	1	1	1	1	77
11	91	19	1	3	3	1	58	60	23	5
12	55	17	3	1	3	1	20	79	10	1
13	95	17	1	7	7	1	40	72	15	87
14	79	15	1	5	7	1	81	10	1	69
15	46	15	3	3	5	3	7	62	66	51
16	63	13	5	5	3	3	80	37	50	56
17	48	15	5	3	7	3	9	53	58	28
18	73	17	3	3	5	3	72	48	60	35
19	38	11	5	1	3	1	27	30	39	36
20	7	7	3	1	1	1	36	63	72	45
21	21	7	1	3	3	1	50	1	1	61

Fig. 10.4 Discretized data for case study-1

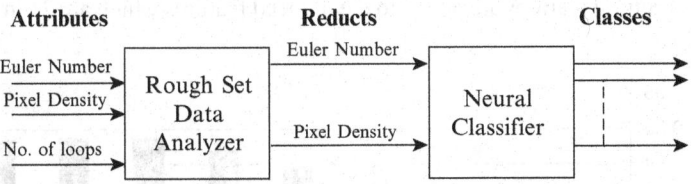

Fig. 10.5 Use of rough set at preprocessing level as first approach

The second approach suggests modification to the architecture of the two layer neural network. In pure neural approach each input layer node is connected with each node of output layer. After computation of reducts we easily know the most significant contributors (input features) to the classification process [11, 12]. Hence depending on the set of reducts obtained the links related to the insignificant features can be removed which further results in lesser computational overheads.

10.6   Results

For both the case studies the training to testing proportion was 80:20 in percentage. The *IS* for case study one has condition attributes of two types structural and statistical which respectively brings consistency and inconsistency to the data. In the second case study the approaches were studied for data of 1,000 user path profiles with different combinations of timestamps and received signal strengths. Hence *IS* for second case study is highly inconsistent.

10.6.1 Results for Case Study-1

The data set generated is given to a pure neural network as described earlier with learning rate $\alpha = 0.4$ and the network converges in almost 1,000 iterations. The attribute set was then reduced by the proposed method and fed into the neural network with the set learning rate. The graphs shown in Fig. 10.6 is achieved for reduction of the attributes space using the above discussed method. The horizontal axis shows the number of resulted reductions and the vertical axis shows the corresponding classification accuracies. From this graph, we conclude that the attribute reductions result in the optimum classification accuracy. Thus the total 31 atributes are reduced to 18 as maximum reduction, 21 as moderate reduction and 26 as minimum reduction. It is found that the minimum classification accuracy of 92.5% is much comparable with the accuracy achieved for pure neural approach using all 31 attributes. The number of reducts to be used hence always is trade off between feature space diamensionality to handled and classification accuracy. This is because too much reduction results in loss of information also. It is observed that at the point of dimentionality reduction certain structural attributes which contribte to the discernability of the system are also neglected (for example numberof loops in the character). Forceful addition of such feature to the reducts yeilds a better efficency. This is shown in Table 10.2. Note that the structural features are added to the reduced features which are 21 in number.

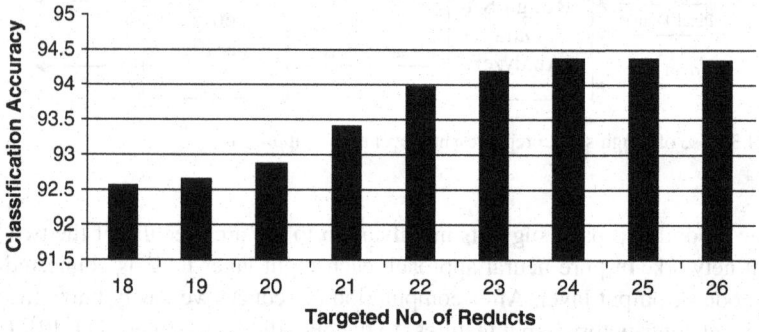

Fig. 10.6 Different classification accuracy values for respective number of reducts

Table 10.2 Classification accuracy for different approaches for case study-1

The kind of data given to the network	Entire attribute space to pure neural network	Reduced attributes from the reducts algorithm (21 No.)	Reduced atributes (21 No.) and loops
Classification accuracy	92.37%	93.76%	94.39%

10.6.2   Results for Case Study-2

For case study two as explained earlier data of 1,000 user path profiles are used. This results in *IS* with seven condition attributes. BST2, BST3, BST4 (base station ids), RSS (received signal strength), TMSTMP (time stamp), SPEED (speed of user motion) and TIME (time). The preprocessing approach results in two sets of reducts with three and four features respectively. Hence for verification of the second approach (rough) neural network with missing links was used. The respective values for classification accuracy are shown in Table 10.3.

Also from the error curves drawn in Figs.10.7–10.9 respectively for various methods, it is seen that pure neural network with reducts and neural network designed with rough set philosophy by missing links, converge much faster than the conventional unsupervised neural network.

Table 10.3 Classification accuracy for different approaches for case study-2

The kind of data given to the network	Entire attribute space to pure neural network	Reduced attributes from the reducts algorithm (3 No.)	Reducts attribute with four missing links in rough neural network
Classification accuracy	98.97%	99.06%	98.23%

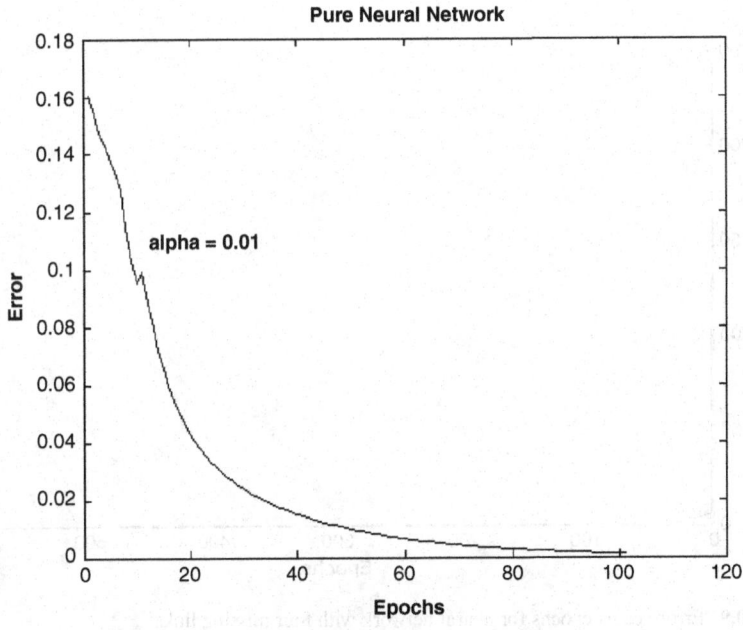

Fig. 10.7 Error versus epochs for pure neural approach for case study-2

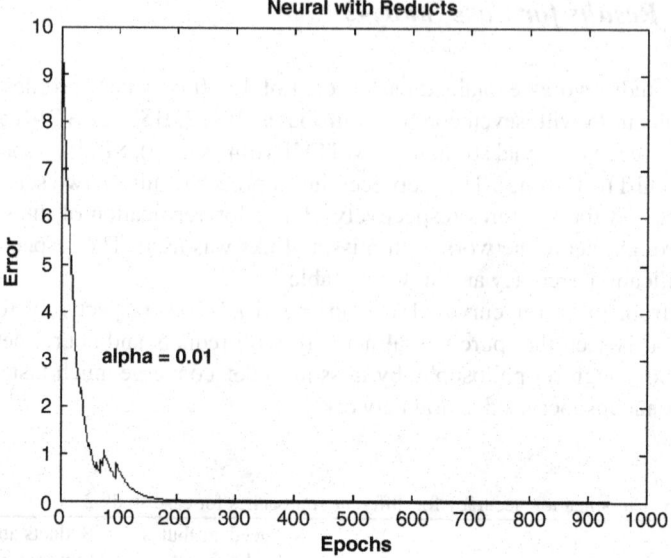

Fig. 10.8 Error versus epochs for neural network fed with reducts for case study-2

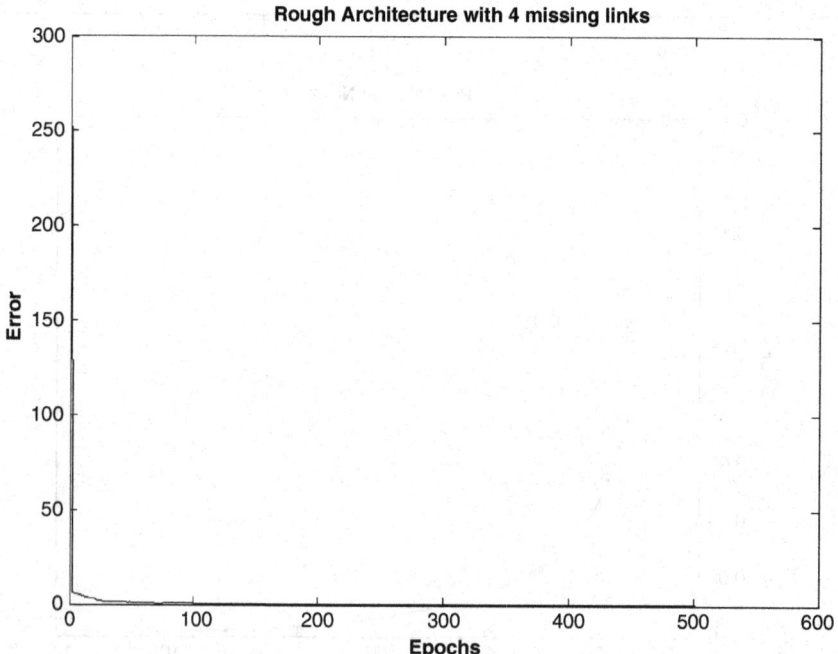

Fig. 10.9 Error versus epochs for neural network with four missing links

10.7   Conclusion

The data set generated contains both the consistent and inconsistent values. According to the rough sets theory consistent attribute values contribute less towards classification but on observation in case study-1, the important structural features such as the loops and crossings when included improves the efficiency of classification. Thus using rough set theory and analysing the importance of features from classification point of view efficient reduction algorithms can be developed. Also it has been observed from the results of case study-2 that for reduct based training or network design we get improved classification accuracy with faster convergence. In future such algorithm can be tested for rough neuron based networks where input data is split in to two parts for specially designed rough neuron.

Acknowledgment The support of Dr. A.P. Gokhale and the team of students consisting of Mr. Bharthan Balaji, Mr. Pradeep Dhananjay, Ms. Y.T. Vasavdatta and Ms. Deepti pant is highly acknowledged for carrying out the experimentation and acquiring of data in the lab.

References

1. Zdzislaw Pawlak: "ROUGH SETS – Theoretical Aspects of Reasoning About Data", 1992, Kluwer, Dordrecht, pages 1–43.
2. Rafael C. Gonzalez, Richard E. Woods, Steven L. Eddiins: "Digital Image Processing Using MATLAB", First Impression, 2006, Pearson Education, NJ, USA, pages 348–497.
3. Jianming Hu, Donggang Yu, Hong Yan: "Algorithm for stroke width compensation of hand-written characters", Electronics Letters Online No. 19961501.
4. Hongsheng Su, Qunzhan Li: "Fuzzy Neural Classifier for Fault Diagnosis of Transformer Based on Rough Sets Theory": IEEE, CS, 2223 to 2227.
5. Giorgos Vamvakas: "Optical Character Recognition for Handwritten Characters": National Center for Scientific Research, Demokritos Athens – Greece, Institute of Informatics and Telecommunications and Computational Intelligence Laboratory (CIL).
6. Jiang-Hong Man: "An Improved Fuzzy Discretization Way for Decision Tables with Continues Attributes", Proceedings of the Sixth International Conference on Machine Learning and Cybernetics, Hong Kong, 19–22 August 2007.
7. S.N. Sivanandam, S. Sumathi, S.N. Deepa: "Introduction to Neural Networks Using Matlab 6.0" first edition, 2006, Tata MCGraw Hill, OH, USA, pages 531–536.
8. J.S.R. Jang, C. T Sun, E. Mizutani: "Neuro-fuzzy and Soft Computing: A Computational Approach to Learning and Machine Intelligence", First edition reprint 2005, Pearson education, NJ, USA, pages 327–331.
9. Xian-Ming Huang, Ji-Kai Yi, Yan-Hong Zhang: "A method of constructing fuzzy neural network based on rough set theory", International Conference on Machine Learning and Cybernetics, 2003, Publication Date: 2–5 Nov. 2003, Volume: 3, pages 1723–1728.
10. Ashwin G. Kothari: "Data Mining Tool for Semiconductor Manufacturing Using Rough Neuro Hybrid approach", Proceedings of International Conference on Computer Aided engineering-CAE-2007, IIT Chennai, 13–15 December 2007.
11. C. Sandeep, R. Mayoraga: "Rough Set Based Neural Network Architecture", International Joint Conference on Neural Networks, Vancouver, BC, Canada, 2006.
12. Pawan Lingras: "Rough Neural Network," Proceedings of the 6th International Conference on Information Processing and Management of Uncertainty, Granada, pages 1445–1450, 1996.

Chapter 11
A New Robust Combined Method for Auto Exposure and Auto White-Balance

Quoc Kien Vuong, Se-Hwan Yun, and Suki Kim

Abstract This paper proposes a new auto-exposure and auto white-balance algorithm that can accurately detect high-contrast lighting conditions and improve the dynamic range of output images for a camera system. The proposed method calculates the difference between the mean value and the median value of the brightness level of captured pictures to estimate lighting conditions. After that, a multiple exposure mechanism which can improve image details is carried out in combination with a simple auto white-balance algorithm which is capable of detecting pictures with one primary color. Simulation results show that the system works well with CMOS sensors used in mobile phones and surveillance cameras. Besides, the proposed algorithm is fast and simple and therefore can be fitted in most CMOS platforms that have limited capabilities.

Keywords Auto exposure · Auto white-balance · Multiple exposure · Mean value · Median value · High-contrast · High dynamic range · Primary color

11.1 Introduction

Auto-exposure (AE) and auto white-balance (AWB) have become two major functions of digital camera systems. Many platforms that provide both AE and AWB controls have been proposed to accommodate various shooting conditions in order to improve the overall system performance.

Many AE algorithms [1–4] have been developed to deal with high-contrast lighting conditions. However, most of these algorithms have some drawbacks on either their accuracy or complexity which may prevent them from being applicable to low-capacity camera platforms such as those employing CMOS technologies. According to Liang et al. [1], it is difficult to discriminate back-lit conditions from

Q.K. Vuong (✉), S.-H. Yun, and S. Kim
ULSI Lab, Korea University, Seoul, Korea
e-mail: quockienvuong@korea.ac.kr

S.-I. Ao et al. (eds.), *Advances in Machine Learning and Data Analysis*,
Lecture Notes in Electrical Engineering 48, DOI 10.1007/978-90-481-3177-8_11,
© Springer Science+Business Media B.V. 2010

front-lit conditions using histogram methods [2,3]. Further simulations in this paper shows that the tables and criteria used to estimate lighting conditions are confusing and not consistent.

Other algorithms [3,4] used fixed-window segmentation methods to estimate the brightness and lighting conditions. Besides, these papers and Ref. [1] only considered images with only one main object. Therefore, these algorithms are not flexible and do not work well with images without a main object.

In the real-time image fusion mechanism proposed by Kao et al. [5], multiple exposure methods were presented to improve the dynamic range of output pictures. Simulation results showed that its algorithm might easily lead to color inconsistency and bad chromatic transitions.

For AWB, the color gains are controlled such that objects which appear white in human eyes are rendered white in the output image. When a white object is illuminated under a low color temperature, it will appear reddish in the captured image. Similarly, it will appear bluish under a high color temperature. Each AWB algorithm consists of two steps. The first step is illumination estimation, and the second step is image compensation.

According to Ref. [6], there are two categories of AWB: global AWB algorithms and local AWB algorithms. Global AWB methods such as gray world assumption, modified gray world assumptions [7,8] may not work well if the picture is dominated by just a few colors. Local AWB methods [9,10] depend on the existence of white objects or human faces in captured images and therefore cannot work with all images.

This paper introduces a new approach to control AE which can be used to determine the degree of contrast lighting employing a simple and quick method which is presented in Sect. 11.2. Section 11.3 describes how to decide if the condition is normal lit, excessive back lit or just a condition with a high dynamic range. Then the algorithm uses a simple multiple exposure mechanism to improve the dynamic range of the output image so that more details can be revealed. This mechanism is carried out in the same time with a refined and simple AWB method based on the idea of gray color points in Ref. [6] with the ability to detect pictures dominated by a few colors. Section 11.4 describes simulation results. Finally, conclusions are given in Section 11.5.

11.2 Auto Exposure Algorithm for Lighting Condition Detection

11.2.1 Lighting Condition Detection

Lighting conditions can be classified as normal lit, excessive back lit or high contrast. To determine the degree of lighting conditions, the proposed method compares the mean value and the median value of the brightness level of the whole image.

The mean brightness level Bl_{mean} is the average brightness level of the whole image. On the other hand, the median value Bl_{med} is the value of the middle item in a sorted array of brightness levels of all pixels in the image.

For a sorted large-size array, if the values of all elements increase or decrease steadily, the difference between the mean and the median values is not significant. An image possessing normal lit condition follows this rule, and thus the difference D_L between Bl_{mean} and Bl_{med} is insubstantial, especially in the cases of normal and under exposure. However, when an image is captured in back-lit or high-contrast lighting conditions and in cases of under and normal exposure, the values of all items increase or decrease abruptly somewhere within the sorted array of brightness levels. In this case, the middle item may have a very large or very small value, depending on the outweighing number of large-value or small-value elements. This leads to a significant difference between the mean and the median values and therefore, Bl_{mean} is very different from Bl_{med}.

In the case of over exposure, the difference varies unpredictably; nevertheless, in normal- and under-exposed images that possess back-lit condition, Bl_{mean} is always larger than Bl_{med} and the difference trend is stable. Figure 11.1 shows the difference between these two values for various cases. Note that D_{thres} is the threshold of the difference value.

Most platforms employing CMOS image sensors (CIS) provide output images in the RGB format. The green component mostly contributes to the luminance of an image. Therefore, to reduce computational complexity, the proposed system uses the green (G) component as the luminance of an image and all steps are performed based on this component.

The use of the G component as the brightness level instead of the luminance component Y (in the YCbCr or YUV formats) can help eliminate the conversion between RGB to other formats and it does not introduce much difference, as proved in Ref. [1] and further clarified in Table 11.1.

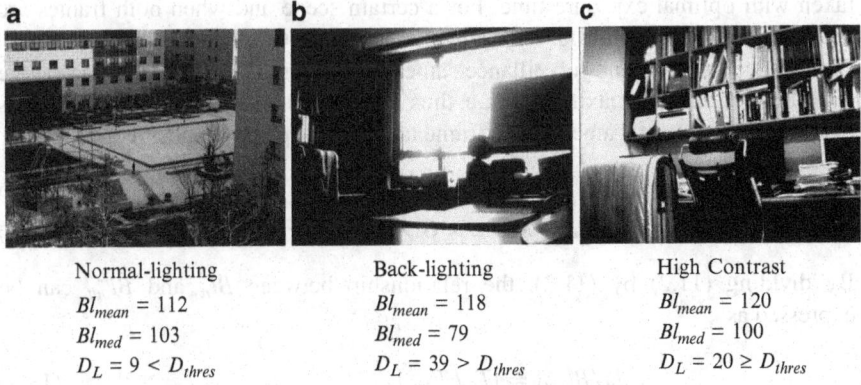

Normal-lighting	Back-lighting	High Contrast
$Bl_{mean} = 112$	$Bl_{mean} = 118$	$Bl_{mean} = 120$
$Bl_{med} = 103$	$Bl_{med} = 79$	$Bl_{med} = 100$
$D_L = 9 < D_{thres}$	$D_L = 39 > D_{thres}$	$D_L = 20 \geq D_{thres}$

Fig. 11.1 Bl_{mean}, Bl_{med} and D_L in different lighting conditions

Table 11.1 G and Y components as brightness level of images in Fig. 11.1

	Bl_{mean}		Bl_{med}		D_L	
Image	G	Y	G	Y	G	Y
(a)	111	112	103	103	8	9
(b)	116	118	76	79	40	39
(c)	120	120	99	100	21	20

In brief, the G component of an RGB image will be used as the luminance when estimating lighting conditions. It is the relationship between the mean and median G values of an image to be used as the criterion to judge illuminating conditions. This relationship will be used in the AE mechanism to help control the exposure value depending on lighting conditions. In term of implementation, the hardware required to compute the mean and median value is simple and among basic blocks. Thus, this method is really effective in terms of processing time and implementation.

11.2.2 Auto Exposure

To address image capturing systems that employ CMOS image sensor and that have limited capabilities, Bl_{mean}, Bl_{med}, D_L and D_{thres} are used to enhance the proposed modified AE algorithm. According to Refs. [1, 11], the relationship between the luminance value and the exposure factors can be expressed as:

$$Bl = k \times L \times G \times T \times (F/\#)^{-2} \tag{11.1}$$

where Bl is the brightness level of the captured image, k is a constant, L is the luminance of the ambient light, G is the gain of the automatic gain control, $F/\#$ is the aperture value, and T is the integration time.

Let Bl_n and Bl_{opt} denote the brightness levels of the current frame and the frame taken with optimal exposure time. For a certain scene and when both frames are taken continuously within a very short time, L and G remain almost the same. For most cell phones and surveillance cameras employing CMOS technologies, the aperture is fixed at its maximum value, thus $F/\#$ is constant. The exposure functions (11.1) for the current frame and the frame taken with optimal exposure time are:

$$Bl_n = k \times L \times G \times T_n \times (F/\#)^{-2} \tag{11.2}$$
$$Bl_{opt} = k \times L \times G \times T_{opt} \times (F/\#)^{-2} \tag{11.3}$$

By dividing (11.2) by (11.3), the relationship between Bl_n and Bl_{opt} can be expressed as:

$$[Bl_n/Bl_{opt}] = [T_n/T_{opt}] \tag{11.4}$$
$$\log_2 Bl_n - \log_2 Bl_{opt} = \log_2 T_n - \log_2 T_{opt} \tag{11.5}$$
$$\log_2 T_{opt} = \log_2 T_n - \log_2 Bl_n + \log_2 Bl_{opt} \tag{11.6}$$

The proposed algorithm uses Bl_{mean} to control AE based on the idea of mid-tone in an iterative way. However, unlike [1] and other methods, in this paper, the optimal brightness level is not fixed. Bl_{opt} may be changed according to the lighting conditions. Besides, since the camera response is not totally linear, the actual values in each condition are obtained by performing a series of experiments. A lot of pictures were taken under different lighting conditions in order to obtain the most suitable optimal values of Bl_{opt} for normal lighting, back lighting or high contrast lighting conditions, and lighting conditions when the current picture is over exposed. These optimal values are expected to be close to the mid-tone value 128, which means that the values of $\log_2 Bl_{opt}$ should be close to $\log_2 128 = 7$.

Let Bl_{opt}^{norm} denote the optimal brightness level in the case of normal-lit conditions with low exposure time, Bl_{opt}^{bkdr} denote the optimal value in the case of back lighting or high contrast lighting conditions with low exposure time, and let Bl_{opt}^{over} denote the optimal value in the case of over exposure.

In real implementation, (11.6) is convenient for data to be stored in look-up tables (LUT). The mid-tone range Bl_{mt} is [100, 130]. After capturing the first frame, the value of Bl_{mean} and Bl_{med} are calculated and are used to decide the value of Bl_{opt} as described in Fig. 11.2. After that, the optimal exposure time is obtained using (11.6). Note that due to the non-linearity of sensors, this mechanism is supposed to be carried out iteratively until Bl_{mean} falls into Bl_{mt}. Different appropriate values of Bl_{opt} help reduce the number of iterations instead of just one common Bl_{opt} for all lighting conditions.

In order to further clarify the exact lighting condition, a multiple exposure algorithm is carried out right after the AE mechanism and in conjunction with the AWB method, as described in the next section.

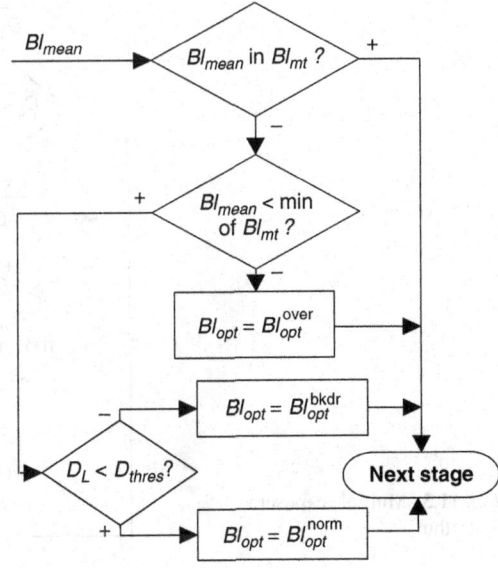

Fig. 11.2 Deciding value for Bl_{opt}

11.3 Multiple Exposure Mechanism and Auto White-Balance

11.3.1 Multiple Exposure Mechanism

After controlling the exposure time so that Bl_{mean} falls into the mid-tone range, the value of Bl_{med} at that optimal exposure level is calculated. Then D_L is obtained as the difference between Bl_{mean} and Bl_{med}.

At this stage, a multiple exposure algorithm described in Fig. 11.3 is employed using two successive frames taken with two different exposure times. The two frames are fused together as follows:

$$F_X(x, y) = \left(F_X^{lo}(x, y) + F_X^{hi}(x, y)\right)/2 \tag{11.7}$$

where $F_X(x, y)$ is the color value of the pixel (x, y), X is either R, G, or B component, lo is low exposure and hi is high exposure.

The multiple exposure mechanism can bring more details to dark areas and over-exposed areas. The frame taken with a lower exposure time provides details; on the other hand, the frame taken with a higher exposure time brightens the fused image.

After image fusion, the updated Bl_{mean} is validated using a modified mid-tone range [90, 130].

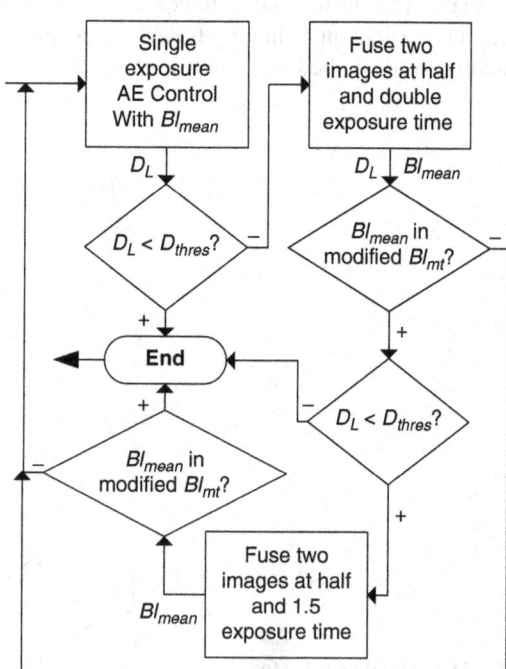

Fig. 11.3 Multiple exposure algorithm

This multiple exposure algorithm is also important to lighting condition estimation since it helps reveal the degree of high contrast lighting as excessive back-lighting, back-lighting, or just high contrast. However, this mechanism is not performed separately but in same time with the AWB method. The AWB is done depending on the actual lighting condition and multiple exposure mechanism is executed differently depending on whether one primary color exists in the picture or not, as discussed in the next section.

11.3.2 Detecting One-Primary Color Images

After AE controlling, lighting conditions are accurately detected. At this stage, the system is already set up with appropriate exposure value.

A lot of pictures were taken to find out common typical characteristics of images that have only a few colors or one primary color. These experiments show that if an image is covered with a primary color, *in most cases, at least one color component (R, G, B) has the median value greater than the mean value.* However, it is interesting to notice that not all colors follow that same rule. Only colors in which either the blue or the red component contributes the most will follow that rule. Other colors that don't satisfy this condition still have all mean values of color components greater than median values. Images with such a primary color may be considered the same as those without any primary color when performing AWB. Figure 11.4a illustrates an image with one primary color but can be considered as other normal images. Figure 11.4b and c illustrate the former kind of images.

An image that has either the median R value greater than the R mean value or the median B value greater than the mean B value is considered as a special one-primary color image. This statement means that any image that satisfies one of the below cases is a *special one-primary color image*:

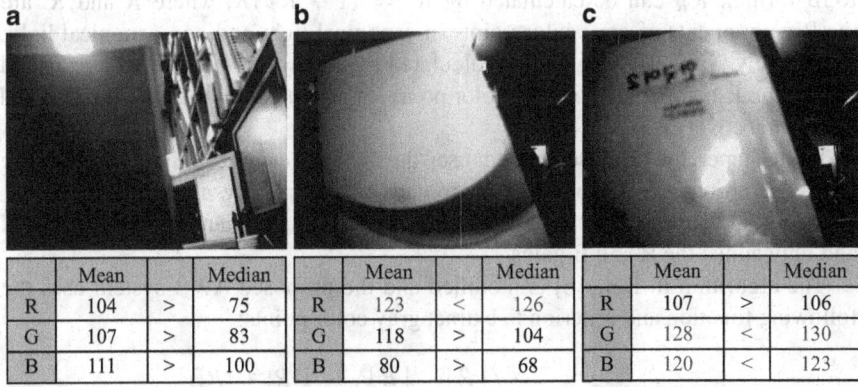

	Mean		Median		Mean		Median		Mean		Median
R	104	>	75	R	123	<	126	R	107	>	106
G	107	>	83	G	118	>	104	G	128	<	130
B	111	>	100	B	80	>	68	B	120	<	123

Fig. 11.4 (a) One-primary color image considered as normal; (b) (c) special one-primary color images

- R: *median* > *mean*; G: median < mean; B: median < mean
- R: *median* > *mean*; G: *median* > *mean*; B: median < mean
- R: median < mean; G: median < mean; B: *median* > *mean*
- R: median < mean; G: *median* > *mean*; B: *median* > *mean*

11.3.3 Auto White-Balance in Conjunction with Multiple Exposure

An AWB algorithm consists of two steps: color temperature estimation, and color channels adjustments. Of the above two steps, the former is more important and it decides the overall accuracy of the whole AWB mechanism. The algorithm in Ref. [6] selects out gray color points from an image to judge the illumination condition. Gray color points can be a white object, a shadow, a black object and so on. Each gray color point has a little deviation from the color gray under different color temperature light sources.

In Ref. [6], the YUV coordinate is used, where Y is the luminance component, and U and V are two chrominance components. Gray color points can be extracted using the following criterion:

$$F(Y, U, V) = \frac{(|U| + |V|)}{Y} < T \qquad (11.8)$$

where T is a threshold value which is far less than 1, and $F(Y, U, V)$ is defined as

$$F(Y, U, V) = (|\tfrac{U}{Y}| + |\tfrac{V}{Y}|) = \frac{(|U|+|V|)}{Y}$$
$$= \begin{cases} \frac{K_R}{1+0.299K_R} \text{in low color temperature} \\ \frac{K_B}{1+0.114K_B} \text{in high color temperature} \end{cases} \qquad (11.9)$$

where K_R and K_B are deviation factors of red (R) and blue (B) components in the RGB format. K_R can be calculated by $R' = (1 + K_R)R$, where R and R' are the R components of gray color points in canonical light and non-canonical light, respectively. Similarly, K_B can be calculated by $B'' = (1 + K_B)B$, where B and B'' are the B components of gray color points in canonical light and non-canonical light, respectively.

In this paper, the proposed system uses the G component to control the exposure time. Therefore, in order to reduce computational complexity, after the AE control, the system will continue to use the G component as the luminance for AWB control and only adjust the R and B gains.

The algorithm in Ref. [6] is modified and the proposed AWB system uses the following function and criterion to extract gray color points:

$$F(R, G, B) = \left(\left|\frac{R}{G}\right| + \left|\frac{B}{G}\right| \right) = \frac{(|R| + |B|)}{G} \qquad (11.10)$$

$$F(R, G, B) = \frac{(|R| + |B|)}{G} < T \qquad (11.11)$$

where T is the threshold value which is far less than 1. However, in the case of pixels whose G value is 0, T is set to 1, and G is set to 1 in (11.10).

Each pixel whose value of $F(R, G, B)$ satisfies (11.11) is accumulated in Ω, where Ω is the set of gray color points.

Let $\overline{R_\Omega}$, $\overline{B_\Omega}$, and $\overline{G_\Omega}$ denote the average of R, B and G components of Ω. These values are used to estimate the illumination condition and to control the R and B gains in an iterative way. Let $\overline{RG_\Omega}$ denote the absolute difference value between $\overline{R_\Omega}$ and $\overline{G_\Omega}$, and let $\overline{BG_\Omega}$ denote the absolute difference value between $\overline{B_\Omega}$ an d$\overline{G_\Omega}$. These values are calculated as:

$$\overline{RG_\Omega} = \left| \overline{R_\Omega} - \overline{G_\Omega} \right| \qquad (11.12)$$

$$\overline{BG_\Omega} = \left| \overline{B_\Omega} - \overline{G_\Omega} \right| \qquad (11.13)$$

When an image is not a special one-primary color image, the AWB control is done if $\left| \overline{RG_\Omega} - \overline{BG_\Omega} \right| \leq D_{AWBthres}$, where $D_{AWBthres}$ is a threshold value; otherwise, if $\overline{RG_\Omega} > \overline{BG_\Omega}$, the R gain needs to be adjusted. Similarly, if $\overline{BG_\Omega} > \overline{RG_\Omega}$, the B gain needs to be adjusted. In this case of normal images, AWB is performed after the multiple exposure mechanism.

When an image is judged as a special one-primary color image, other AWB conditions are used. Furthermore, in the case of back lighting conditions, multiple exposure should not be performed since image fusion may disrupt the color consistency; on the other hand, in the cases of high contrast lighting and normal lighting, multiple exposure can still be carried out to improve the details of output

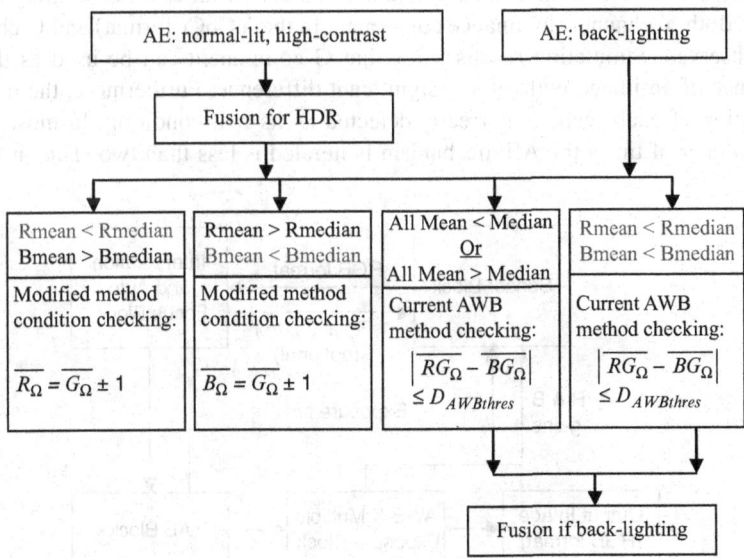

Fig. 11.5 AWB in combination with multiple exposure

images without degrading the chromatic visuality. The detailed combined algorithm for AWB and multiple exposure as well as the AWB conditions for the case of special one-primary color images are described in Fig. 11.5.

11.4 Simulations and Comparisons

Figure 11.6 describes the simplified functional block diagram of the proposed system with AE and AWB functions. The output data of the auto focus (AF) and interpolation block are fed to the AE block. Note that the AF capability is optional. Most CMOS platforms are not equipped with AF function. In the AE block, after multiple exposure controlling, output data are sent to the AWB block.

Simulations were carried out using a simple platform employing CMOS image sensors (CIS) with AE parameter values as follows:

$$D_{thres} = 20; \log_2 Bl_{opt}^{norm} = 6.8; \log_2 Bl_{opt}^{bkdr} = 7; \log_2 Bl_{opt}^{over} = 6.36;$$

Simulation results show that the proposed AE algorithm can detect lighting conditions accurately and does not require much computation (Fig. 11.7). Furthermore, the algorithm is independent from the position of the light source and can work well with images with or without a main object.

Because of the non-linear characteristics of CMOS sensors, sometimes it requires that the AE algorithm be iterated more than once since the first calculated exposure value does not return a value in the range of Bl_{mean} in Bl_{mt}. Therefore the overall AE mechanism may include more than one adjusting time.

Tables 11.2–11.4 demonstrate simulation results for all cases of lighting conditions. Both Y channel (luminance component in the YCbCr format) and G channel are observed. Simulation results show that G component can be used as the luminance of an image without any significant difference. Furthermore, the lighting condition of each scene is correctly detected as its real condition. In most cases, the number of times the AE mechanism is iterated is less than two. This indicates

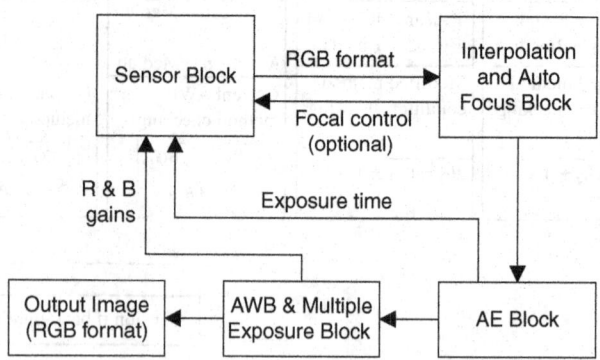

Fig. 11.6 Simplified functional block diagram of the proposed camera system

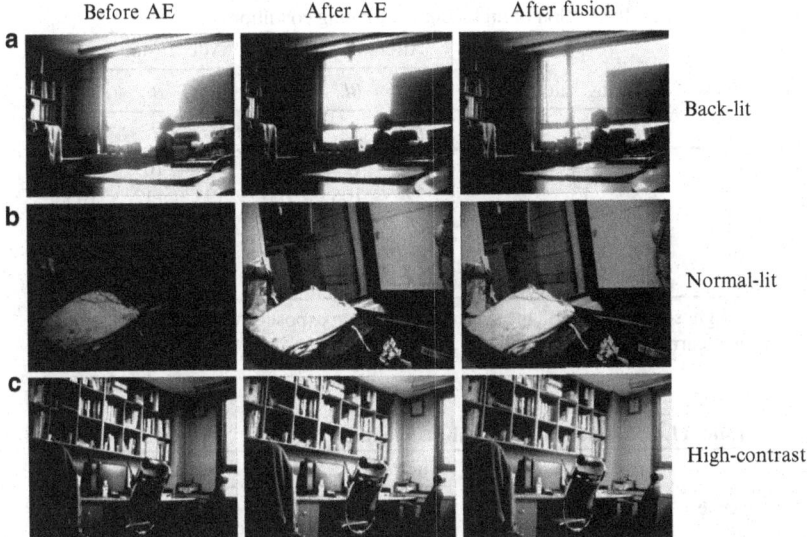

Before AE　　　　　　After AE　　　　　　After fusion

Back-lit

Normal-lit

High-contrast

Fig. 11.7 Simulation with AE algorithm

Table 11.2 Evaluation of back-lit conditions

Scene	Starting values			After AE Bl_n			After fusion Bl_n		
	Bl_n	D_L	Times	D_L	Y	G	D_L	Y	G
(1)	156	8	1	40	118	116	27	123	122
(2)	130	27	1	42	107	104	29	115	112
(3)	160	−6	1	39	121	121	22	121	120
(4)	173	−78	2	39	111	111	24	114	114
(5)	87	49	1	45	115	114	31	119	117

that the proposed algorithm provides a high accuracy rate and fastens the overall performance.

Tables 11.2 and 11.3 describe simulation results of back-lit and high dynamic range (HDR) conditions. The values of D_L after AE controlling and after fusion show that fused images provide more details than un-fused ones. This ability is very useful for camera systems that employ CMOS image sensors with limited dynamic range. In the case of HDR conditions, after AE controlling, the multiple exposure mechanism is carried out twice.

Table 11.4 describes simulation results of images taken in normal-lit conditions. The simulation also shows further values of these pictures after fusing using two images taken at half and 1.5 times the optimal exposure time. These experiment results indicate that this image fusion mechanism can also provide more details in output images for surveillance systems.

Table 11.3 Evaluation of high-contrast lighting conditions

Scene	Starting values			After AE			After fusion		
				Bl_n			Bl_n		
	Bl_n	D_L	Times	D_L	Y	G	D_L	Y	G
(1)	84	22	2	21	120	120	12	109	109
(2)	22	13	2	32	106	100	19	112	105
(3)	77	29	2	25	115	114	13	107	106
(4)	169	−33	2	30	117	116	19	111	111
(5)[a]	37	15	1	45	121	112			

[a] Night scene taken with the system's maximum exposure value; thus no fusion was carried out after AE.

Table 11.4 Evaluation of normal-lit conditions

Scene	Starting values			After AE			After fusion		
				Bl_n			Bl_n		
	Bl_n	D_L	Times	D_L	Y	G	D_L	Y	G
(1)	79	−3	1	−11	117	115	−14	110	109
(2)	82	14	1	14	105	104	8	99	99
(3)	8	3	3	15	109	106	8	99	98
(4)	40	11	1	15	107	111	9	101	104
(5)[a]	3	1	1	0	42	39			

[a] Night scene taken with the system's maximum exposure value.

For AWB control, the value of $D_{AWBthres}$ is set to 1. In Ref. [6], there are three values for T: 0.97, 0.1321, and 0.2753. However, due to limited capabilities and accuracy of CMOS platforms, the threshold value T in the proposed system is set to 0.28. Figure 11.8 demonstrates the evaluation of the whole system including both AE and AWB functions. Figure 11.8d shows that for special one-primary color images, if the method to detect and the corresponding AWB conditions are applied, the actual color of the main object can be revealed exactly. Figure 11.8e illustrates the advantage of using the image fusion mechanism even in the case of high-contrast lighting conditions in term of image details.

11.5 Conclusions

A new AE algorithm with lighting condition detecting capability has been introduced. The proposed algorithm can quickly estimate an appropriate exposure value after a small number of frames. It can also improve the accuracy and enhance the details of output images.

The proposed AWB method, which is performed after AE control, is a simple local AWB algorithm. This algorithm not only provides high accuracy and flexibility

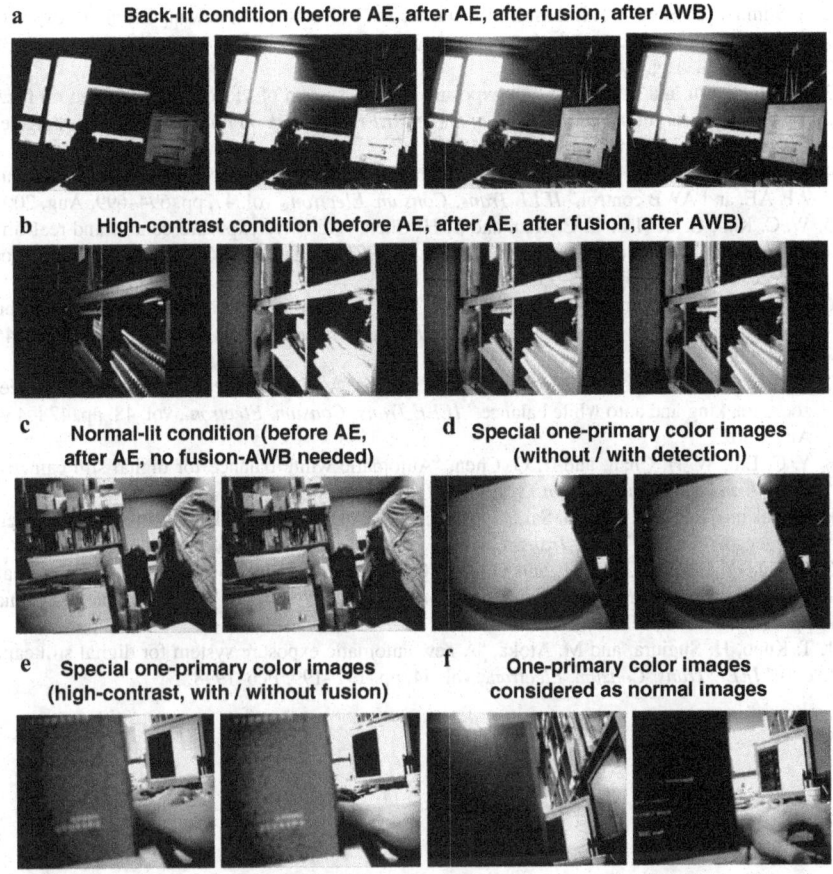

a   **Back-lit condition (before AE, after AE, after fusion, after AWB)**

b   **High-contrast condition (before AE, after AE, after fusion, after AWB)**

c   **Normal-lit condition (before AE,**   **d** **Special one-primary color images**
       after AE, no fusion-AWB needed)   **(without / with detection)**

e   **Special one-primary color images**   **f**   **One-primary color images**
       (high-contrast, with / without fusion)   **considered as normal images**

Fig. 11.8 Simulations with AE and AWB

owing to the omnipresence of gray color points but also is capable of detecting images of one primary color or images dominated with only a few colors.

Using the new mechanism to detect light conditions, the system is flexible and can work well with most images without being affected by the positions of light sources and main objects. Since the algorithm is not computationally complicated, it can be fitted in most CMOS platforms that have limited capabilities such as cell phones and/or surveillance cameras.

References

1. J. Y. Liang, Y. J. Qin, and Z. L. Hong, "An auto-exposure algorithm for detecting high contrast lighting conditions," *Proc. of the 7th Int. Conf. on ASIC*, Guilin, Peoples R. China, vols. 1 and 2, pp. 725–728, Oct. 2007.

2. S. Shimizu, T. Kondo, T. Kohashi, M. Tsuruta, and T. Komuro, "A new algorithm for exposure control based on fuzzy logic for video cameras," *IEEE Trans. Consum. Electron.*, vol. 38, pp. 617–623, Aug. 1992.

3. M. Murakami and N. Honda, "An exposure control system of video cameras based on fuzzy logic using color information," *Proc. of the 5th IEEE Int. Conf. on Fuzzy Systems*, Los Angeles, CA, vols. 1–3, pp. 2181–2187, Sept. 1996.

4. J. S. Lee, Y. Y. Jung, B. S. Kim, and S. J. Ko, "An advanced video camera system with robust AF, AE, and AWB control," *IEEE Trans. Consum. Electron.*, vol. 47, pp. 694–699, Aug. 2001.

5. W. C. Kao, C. C. Hsu, C. C. Kao, and S. H. Chen, "Adaptive exposure control and real-time image fusion for surveillance systems," *Proc. of IEEE Int. Symp. on Circuits and Systems*, Kos, Greece, vols. 1–11, pp. 935–938, May 2006.

6. J. Y. Huo, Y. L. Chang, J. Wang, and X. X. Wei, "Robust automatic white balance algorithm using gray color points in images," *IEEE Trans. Consum. Electron.*, vol. 52, pp. 541–546, May 2006.

7. Y. Kim, J. S. Lee, A. W. Morales, and S. J. Ko, "A video camera system with enhanced zoom tracking and auto white balance," *IEEE Trans. Consum. Electron.*, vol. 48, pp. 428–434, Aug. 2002.

8. Y. C. Liu, W. H. Chan, and Y. Q. Chen, "Automatic white balance for digital still camera," *IEEE Trans. Consum. Electron.*, vol. 41, pp. 460–466, Aug. 1995.

9. N. Nakano, R. Nishimura, H. Sai, A. Nishizawa, and H. Komatsu, "Digital still camera system for megapixel CCD," *IEEE Trans. Consum. Electron.*, vol. 44, pp. 581–586, Aug. 1998.

10. B. Hu, Q. Lin, X. L. Kang, and G. M. Chen, "A new algorithm for automatic white balance with priori," *IEEE Asia-Pacific Conf. on Circuits and Systems*, Tianjin, Peoples R. China, pp. 109–112, Dec. 2000.

11. T. Kuno, H. Sugiura, and M. Atoka, "A new automatic exposure system for digital still cameras," *IEEE Trans. Consum. Electron.*, vol. 44, pp. 192–199, Feb. 1998.

Chapter 12
A Mathematical Analysis Around Capacitive Characteristics of the Current of CSCT: Optimum Utilization of Capacitors of Harmonic Filters

Mohammad Golkhah and Mohammad Tavakoli Bina

Abstract A new shunt reactive power compensator, CSCT, is presented and introduced in this paper. Mathematical analysis of harmonic content of the current of CSCT is performed and use of a winding with additional circuit has been presented as a solution to suppress these harmonics.

Keywords CSCT · Harmonic filter · Reactive power compensation · Thyristor controlled transformer

12.1 Introduction

CSCT stands for "Controlled Shunt Compensator of Transformer type" [1–4]. A general scheme of this compensator is presented in Fig. 12.1. This configuration is a transformer with three windings. NW is network winding which is connected to the network and is the main winding of the compensator. CW is the second winding to which a thyristor valve and a parallel voltage circuit breaker are connected and is called CW briefly. The third winding is compensating winding which is indicated by ComW in Fig. 12.1. Two highest harmonic filters and a capacitor bank are connected to this winding. It is important to note that CSCT is a three phase compensator. The connection of NWs of three phases is star and the neutral is grounded. Control windings' connection is as same as network windings of three phases. However compensating windings can be delta in connection together.

When the thyristor is opened all the magnetic flux passes through the magnetic core leading to a minimum reluctance, maximum inductance and a capacitive current in NW according to the value of capacitor bank and eventually generate reactive power to the network. On contrary, when the thyristor is closed the flux is subjected to pass through air gap including all of the windings. Hence the reluctance, inductance and the current of NW will be maximal, minimal and maximal inductive (the rated value) respectively.

M. Golkhah (✉) and M.T. Bina
Electrical Engineering Department, K. N. Toosi University of Technology, Tehran, Iran
e-mail: goulkhah_elect@yahoo.co.uk

S.-I. Ao et al. (eds.), *Advances in Machine Learning and Data Analysis*,
Lecture Notes in Electrical Engineering 48, DOI 10.1007/978-90-481-3177-8_12,
© Springer Science+Business Media B.V. 2010

Fig. 12.1 General scheme of
a CSCT: TV, thyristors valve;
VCB, vacuum circuit breaker;
C5-L5 and C7-L7 filters of
fifth and seventh harmonics;
C, additional capacitor bank

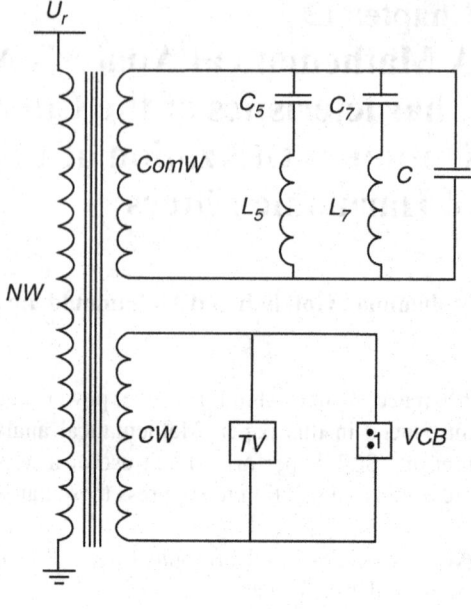

Fig. 12.2 Mono phase
diagram of CSCT: **a**, view
form side; **b**, view form
above: 1, a main core; 2,
yokes; 3, lateral yokes; 4,
NW; 5, CW; 6, ComW; 7,
magnetic shunts

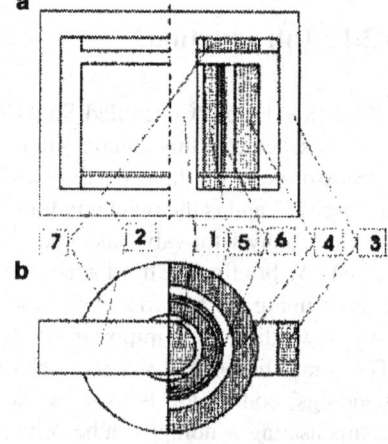

Since two modes lead to two different signs in the reactive power, we can claim that the compensator is bilateral. However the capacitor can be removed from the compensating winding still having bilateral reactive power compensation if the relationships of designing elements of harmonic filters be correctly performed. These relations are calculated and presented below.

Figure 12.2 represents one phase of the transformer with mentioned three windings. The winding close to the main core is CW, the outer winding is NW and interlayed winding is ComW.

12.2 Mathematical Analysis

Flowing capacitive current in the network winding of CSCT under open circuit condition of CW, which defines the capacitive conductance of harmonic filters of fifth and seventh in the main frequency, has prompted the possibility of bilateral action of the compensator with increased capacitive component of filters of the higher harmonics (see Fig. 12.3).

Impedance of the compensator when thyristors are completely closed is defined by (12.1).

$$
\begin{aligned}
X_{Leq} &= X_1 + \frac{X_2 X_3}{X_2 + X_3} = \delta X_{12} + \frac{(1 - \delta) X_{12} X_3}{(1 - \delta) X_{12} + X_3} \\
&= X_{12} \left[\delta + \frac{(1 - \delta)}{1 + \frac{X_{12}}{X_3} (1 - \delta)} \right] > X_{12}
\end{aligned}
\tag{12.1}
$$

This value is greater than the corresponding impedance of controlled shunt *reactor* of transformer type, CSRT.

$$
\begin{aligned}
X_{L.eq.CSRT} &= X_1 + \frac{X_2 X_{f.eq}}{X_2 + X_{f.eq}} = \delta X_{12} + \frac{(1 - \delta) X_{12} X_{f.eq}}{(1 - \delta) X_{12} + X_{f.eq}} \\
&= X_{12} \left[\delta + \frac{(1 - \delta)}{1 + \frac{X_{12}}{X_{f.eq}} (1 - \delta)} \right] \approx X_{12}
\end{aligned}
\tag{12.2}
$$

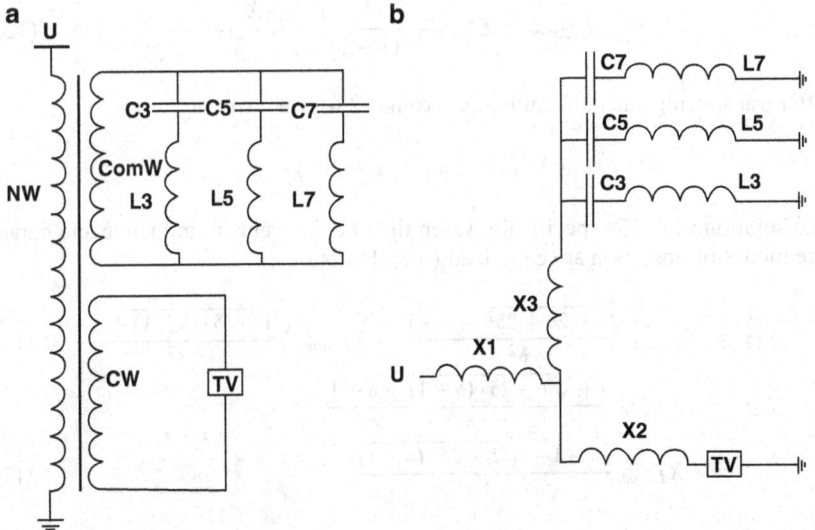

Fig. 12.3 Schematic single-line diagram (**a**) and equivalent circuit (**b**) of CSCT without any capacitor bank

To provide the rated current by CSCT as same as CSRT, the impedances X_{12}, X_{13} and X_{23} should be lowered. For instance this objective can be achieved by reducing the number of turns in windings or increasing the height of magnetic window.

In a general view, the solution to this problem can be carried out on the basis of two equations made for locked and opened thyristors. In open thyristors condition, we want the capacitive current flowing through the network winding of CSCT to be equal to its rated current, then

$$I_c = \frac{U_{ph}}{8X_{12} + X_c} = -\alpha \frac{U_{ph}}{X_{L.nom}} \tag{12.3}$$

Where X_{Lnom} is the essential rated impedance of compensator in inductive mode of operation and α is required correlation between rated capacitive and inductive currents of the compensator.

The essential value of X_C is obtained from (12.4).

$$X_c = \frac{1}{\alpha} (X_{L.nom} + \alpha\delta \cdot X_{12}) \tag{12.4}$$

Let's now suppose that the inductive current through network winding of the compensator is equal to its rated current with completely closed thyristors:

$$I_L = \frac{U_{ph}}{8X_{12} + \frac{(1-\delta)X_{12}X_c}{(1-\delta)X_{12}+X_c}} = \frac{U_{ph}}{X_{L.nom}} \tag{12.5}$$

Then we have from (12.5):

$$X_{L.nom} = 8X_{12} + \frac{(1-\delta) X_{12}X_c}{(1-\delta) X_{12} + X_c} \tag{12.6}$$

After transposing and substituting Xc from (12.4), we have:

$$\alpha\delta^2 X_{12}^2 + X_{12}X_{Lnom} (1 + \alpha - 2\alpha\delta) - X_{L.nom}^2 = 0 \tag{12.7}$$

The solution of (12.7) specifically when the rated currents in inductive and capacitive modes of operation are equalized ($\alpha = 1$) obtains:

$$X_{12} = X_{L.nom} \frac{\sqrt{1 - 2\delta + 2\delta^2} + \delta - 1}{\delta^2} = X_{L.nom} \frac{\sqrt{1 - 2\delta \cdot (\delta - 1)} + \delta - 1}{\delta^2} \tag{12.8}$$

$$X_C = -X_{L.nom} \frac{\delta + \sqrt{1 - 2\delta \cdot (\delta - 1)} + \delta - 1}{\delta}$$

$$= -X_{L.nom} \frac{2\delta - 1 + \sqrt{1 - 2\delta \cdot (\delta - 1)}}{\delta} \tag{12.9}$$

For example when $\delta = 0.6$

Fig. 12.4 Dependences of relative value of reactance between NW and CW windings on relative value of reactance between NW and ComW $\delta = X_{13}/X_{12}$ at different ratios of rated currents in distribution and consumption conditions of reactance: $1 - \alpha = 0.2$; 2–0.4; 3–0.6; 4–0.8; 5–1.0

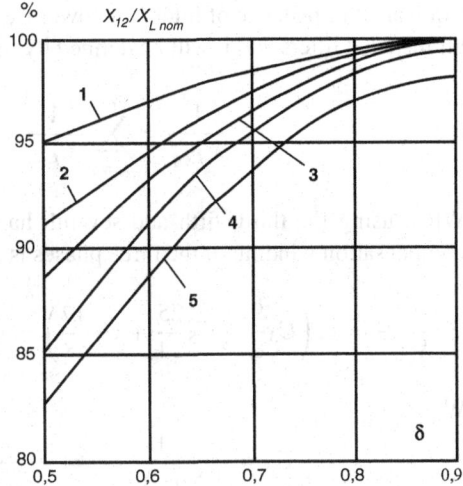

$$X_{12} = X_{L.nom}\frac{\sqrt{0.52} - 0.4}{0.36} = 0.891 X_{L.nom} \text{ and}$$

$$X_c = -1.535 X_{L.nom} = -1.72 X_{12}$$

Since δ varies between 0.5–0.8, the rated inductance $X_{L.nom}$ is greater than X_{12} under any value of α. Hence the compensator's impedance under closed CW condition without filters is less than 100% unlike from CSRT (see Fig. 12.4).

It means that under the same rated inductive current and equal dimensions, which define X_{12}, the number of turns in windings of the compensator is less than that of CSRT. The ratio $X_{12}/X_{L.nom}$ is increased by increasing δ, and decreased when α is increased. Hence, higher is the value of α, smaller is the number of turns in the windings of CSCT in comparison with CSRT with the same capacity.

Reduction in the number of coils leads to increment in magnetic flux when thyristors are locked and when completely opened. Hence, in CSCT the active cross-section of magnetic conductor steel is greater than that of CSRT with the same capacity.

Further, we shall define the essential capacitance of highest harmonics filters to provide the relation α from the obtained value of Xc and known ratios for filters. The equivalent impedance of filters is defined by the ratio:

$$X_1 + X_{f.eq} = -\frac{X_{L.nom}}{\alpha}$$

So we have:

$$X_{f.eq} = \frac{X_{L.nom}}{\alpha}\left[1 + \frac{\sqrt{(1 + \alpha - 2\alpha\delta)^2 + 4\alpha\delta} - (1 + \alpha - 2\alpha\delta)}{2\delta}\right] \tag{12.10}$$

Equivalent impedance of filters on power frequency is defined by the impedance of all installed filters and it will be defined by the ratio:

$$\frac{1}{X_{f.eq}} = \sum_{k=3}^{n} \frac{1}{X_{f.eq}} = \sum_{k=3}^{n} \frac{\omega \cdot C_k k^2}{1 - k^2} \tag{12.11}$$

When using the third, fifth and seventh harmonic filters (if for some reason the compensation winding of the three phases is not in delta connection), we obtain:

$$\frac{1}{X_{f.eq}} = \omega \left(C_3 \frac{9}{8} + C_5 \frac{25}{24} + C_7 \frac{49}{48} \right) = \omega C_3 \left(1125 + 1.04 \frac{C_5}{C_7} \right) = 1.5\omega C_3 \tag{12.12}$$

Whence

$$C_3 = \frac{1}{1.5\omega \cdot X_{f.eq}} = \frac{\alpha}{1.5\omega X_{L.nom} \left(1 + \frac{\alpha\delta \cdot X_{12}}{X_{L.nom}} \right)} \tag{12.13}$$

and C5 = 0.26C3; C7 = 0.105C3.

Voltage of capacitors of filters equals:

$$\Delta U_{c.f} = 1.1 U_{ph} \left[1 + \frac{\sqrt{(1 + \alpha - 2\alpha\delta)^2 + 4\alpha\delta^2} - (1 + \alpha - 2\alpha\delta)}{2\delta} \right] \tag{12.14}$$

and the ratio of capacitor power of filters to the rated capacitive power of capacitors is:

$$\frac{Q_{c.f}}{Q_{C.CSCT}} = 1.1 \left[1 + \frac{\sqrt{(1 + \alpha - 2\alpha\delta)^2 + 4\alpha\delta^2} - (1 + \alpha - 2\alpha\delta)}{2\delta} \right] \tag{12.15}$$

Corresponding dependences of ratios $Q_{c.f}/Q_{c.csct}$ are shown in Fig. 12.5.

Exceeding the voltage of compensating winding in comparison with the voltage defined by the transformation ratio leads to increment of the magnetic flux in the core. In order to eliminate the saturation of the core, it is essential to increase the active cross-section of the core proportional to the voltage increase (Fig. 12.6).

When the thyristors are opened, the current in branch 2 (Fig. 12.3) equals to zero and the voltage in the thyristors equals to the voltage in ComW (certainly in view of the ratio of the number of turns in CW and ComW ($K_{T.2-3} = K_{T.1-3}/K_{T.1-2}$). Consequently, with equal number of turns in ComW and CW, rated voltage of CW is defined by formula 12.14. With difference in number of turns in ComW and CW, the rated voltage in CW

$$U_{2nom} = \Delta U_{C.f} K_{T.2-3}$$

Fig. 12.5 Dependences of amount of third, fifth and seventh harmonics capacitor power – rated capacitive power of CSCT ratio on the relative values of inductive resistance of CSCT between ComW and CW $\delta = X_{13}/X_{12}$ at different values of $\alpha = Q_{c.f}/Q_{L.CSCT}$. ComW of the three phases are in Y-connection with neutral terminal

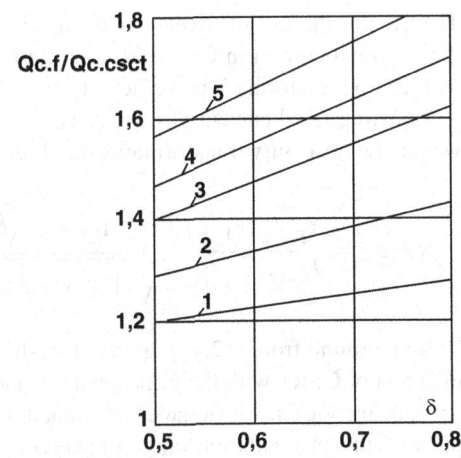

Fig. 12.6 Dependence of maximum current of CW on α when $\delta = 0.5$

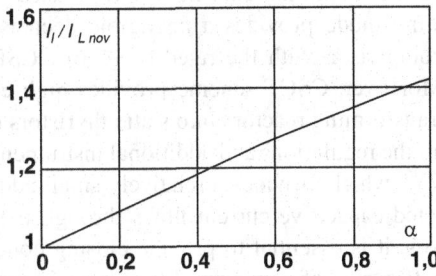

The rated voltage of CW is chosen based on the appropriate (by technical and economic considerations) current through the thyristors. We shall obtain the relationship between rated current of CW and voltage in NW (see Fig. 12.3):

$$I_{2.nom}X_2 = U_{ph} - I_{L.nom}X_1$$

or

$$I_{2nom} = \frac{\delta}{1-\delta}I_{Lnom} \cdot \frac{1+\alpha - \sqrt{(1+\alpha-2\alpha\delta)^2 + 4\alpha\delta^2}}{\sqrt{(1+\alpha-2\alpha\delta)^2 + 4\alpha\delta^2} - (1+\alpha-2\alpha\delta)} \tag{12.16}$$

For example when $\delta = 0.5$ according to (12.16)

$$\frac{I_{2nom}}{I_{Lnom}} = \sqrt{1+\alpha}$$

The greater the ratio between rated capacitive and inductive currents α, the more the maximal current in CW increases relative to the rated inductive current in NW (subject to transformation coefficient).

Equating rated current of CW to the chosen rated current of thyristors, IT.nom, we get the necessary transformation coefficient:

$$K_{T.1-2} = \frac{I_{T.nom}}{I_{L.nom}} \cdot \frac{1-\delta}{\delta} \cdot \frac{1 + \alpha - \sqrt{(1 + \alpha - 2\alpha\delta)^2 + 4\alpha\delta^2}}{\sqrt{(1 + \alpha - 2\alpha\delta)^2 + 4\alpha\delta^2} - (1 + \alpha - 2\alpha\delta)} \quad (12.17)$$

It can be found from (12.17) that the transformation coefficient of CSCT is smaller than that of CSRT with the same capacity and it depends on α and δ.

Thus increase in capacitance of capacitors enables to essentially expand the capacitor current regulation range of the compensator towards inductive current up to $\pm 100\%$ of rated CSRT current. Hence one thyristors block of CSRT calculated on the maximum current in CW is used. Quick closing of CW in nominal operating mode provides a negligible decrease in the rated current of the CSCT in comparison with the rated current of CSRT without supplementary capacitance. Moreover, CSCT scheme provides high efficient use of capacitance of filters by transforming reactors into static thyristors compensators. It is essential to note that all the regulations and additional instruments are carried out in the low voltage side, CW, which provides a relatively small additional costa in relation to CSRT. Since rated capacitive current flows through the network winding under locked thyristors, it is essential to provide closing conditions of thyristors in some part of the half-period of commercial frequency in order to reduce it. At a particular combustion angle of thyristors, the inductive current matches with the capacitive current and the network winding current reaches to zero. At further increase in combustion angle of thyristors, current becomes inductive and increases right up to the rated value when the combustion angle is $180°$ (complete conducting of thyristors). At zero current through the network winding, equivalent impedances of limbs 2 and 3 of CSCT three-beam schemes are the same. Hence equivalent impedance of thyristors limbs is equal to:

$$X_{2eq} = \frac{X_{Lnom}}{2\delta\alpha} \left[2\delta + \sqrt{(1 + \alpha - 2\alpha\delta)^2 + 4\alpha\delta^2} - (1 + \alpha - 2\alpha\delta) \right] \quad (12.18)$$

Since at zero current in the network winding of the compensator, voltage loss in inductive impedance $X_1 = \delta X_{12}$ equals to all voltages applied to any other two limbs of Fig. 12.3. Thus current in the thyristors limb equals to:

$$I_{2.0} = \frac{U_{ph}}{X_{2.eq}} = \frac{2\alpha\delta U_{ph}}{X_{L.nom}} \frac{1}{2\delta + \sqrt{(1 + \alpha - 2\alpha\delta)^2 + 4\alpha\delta^2} - (1 + \alpha - 2\alpha\delta)}$$
$$(12.19)$$

Fig. 12.7 Dependence on α of relative value of current through thyristor block when current in the network winding passes through zero (*curves 1,2*) at rated phase voltage (*curve 1*) and at voltage in CW under completely closed thyristors (*curve 2*), as well as the relative value of that voltage (*curve 3*)

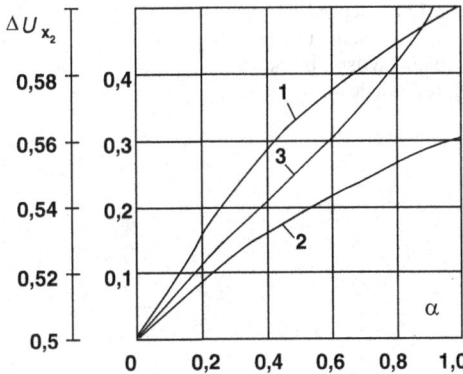

And accordingly the relative value of current in thyristor limb under zero current in NW considering (12.16) will be:

$$\frac{I_{2.0}}{I_{2.nom}} = 2\alpha\,(1-\delta)\,\frac{2\delta + \sqrt{(1+\alpha-2\alpha\delta)^2 + 4\alpha\delta^2} - (1+\alpha-2\alpha\delta)}{1+\alpha - \sqrt{(1+\alpha-2\alpha\delta)^2 + 4\alpha\delta^2}} \times$$

$$\times \frac{1}{2\delta + \sqrt{(1+\alpha-2\alpha\delta)^2 + 4\alpha\delta^2} - (1+\alpha-2\alpha\delta)} \qquad (12.20)$$

As seen, when the current passes through zero, the relative value of current in thyristor limb is defined by only two variables: α and δ. However calculations show that the ratio I2/I2nom does not depend on δ when I1 = 0 and its dependence on α is shown in Fig. 12.7, which is an approximated function (with inaccuracy not more the 2%).

$$I_2/I_{2nom} = 0.91\alpha \cdot e^{-0.6\alpha} \qquad (12.21)$$

The ignition angle and correspondingly the combustion angle of thyristors when CSCT current passes through zero by the corresponding thyristor characteristic (Fig. 12.8) can define the relative value of current. However it follows to bear in mind that characteristic Fig. 12.8 was obtained at constant voltage and constant inductive impedance connected in series with thyristors. In the case of CSCT, inductive impedance $(1 - \delta)X_{12}$ remains constant when any current passes through the thyristors and voltage at increased current increases right up to phase voltage at equal reactance of inductive and capacitive limbs.

In nominal inductive operating mode of CSCT, the current through inductance $X_2 = (1 - \delta)X_{12}$ is defined by formula (12.16), from where voltage drop in that inductive resistance is:

$$\Delta U_{X_2} = L_{2nom}X_{12}\,(1-\delta) = \frac{U_{ph}}{2\alpha\delta}\left[1+\alpha-\sqrt{(1+\alpha-2\alpha\delta)^2 + 4\alpha\delta^2}\right] \qquad (12.22)$$

Fig. 12.8 Dependence of relative value of current through thyristor block on ignition angle of thyristors

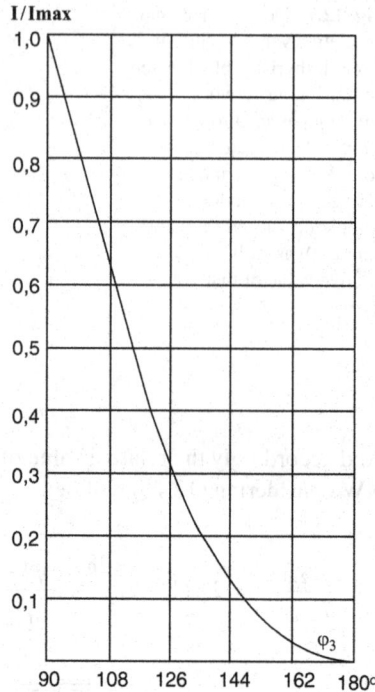

and consequently the ratio of the voltage drop in the thyristor to the phase voltage changes with change in α (see Fig. 12.7), although this change is negligible (in the range of 18%).

In the Fig. 12.7, the dependence of $f(\alpha)$ brought to rated voltage in CW (when thyristors are completely opened) by multiplying the given curve 1 and the corresponding ratio of the voltage drop in the thyristor to the phase voltage (curve 3). This dependence is well approximated by below formula.

$$\frac{I_{2.0}}{I_{2.nom}} = 0.43\alpha \cdot e^{-0.46\alpha} \tag{12.23}$$

This last dependence in conjunction with the curve of Fig. 12.8 enables us to define the dependence of ignition angle and combustion angle of thyristors when network winding current passes through zero on the value of α:

$$
\begin{array}{cccccc}
0.043 & 0.2 & 0.4 & 0.6 & 0.8 & 1.0 \\
\underline{168^0} & 155° & 143° & 135° & 131.4° & 129° \\
\underline{24^0} & 50° & 74° & 90° & 97.2° & 102°
\end{array}
$$

Thus with increase in α, the ignition angle of thyristors when CSCT current passes through zero decreases and combustion angle of thyristors increases correspondingly.

The obtained result enables us to change the definition of angle characteristics of CSCT: dependence of CSCT current on the ignition angle of thyristors.

Let us bring the relative reactance of thyristor limb K in the earlier stated formulas of CSCT taking the base value of reactance when CSCT current passes through zero:

$$X_{2eq} = -KX_C = \frac{K}{\alpha}(X_{Lnom} + \alpha\delta X_{12}) \qquad (12.24)$$

Thus we get the equivalent reactance of CSCT at any value of K in the form of:

$$X_{eq} = X_{Lnom}\frac{2K\sqrt{(1+\alpha-2\alpha\delta)^2 + 4\alpha\delta^2} - (1+\alpha-2\alpha\delta)}{2(1-K)\alpha\delta} \qquad (12.25)$$

Consequently, the ratio of CSCT current to rated inductive current at any value of K

$$\frac{I_{(K)}}{I_{L.nom}} = \frac{X_{Lnom}}{X_{eq}} = \frac{2(1-K)\alpha\delta}{28 \cdot K + \sqrt{(1+\alpha-2\alpha\delta)^2 + 4\alpha\delta^2} - (1+\alpha-2\alpha\delta)} \qquad (12.26)$$

Considering that the value δ does not influence the relative value of current, we take it to equal its minimal value $\delta = 0.5$.

Fig. 12.9 Dependence of the relative value of current of CSCT on ignition angle of thyristors: $1 - \alpha = 0.043$; 2–0.2; 3–0.4; 4–0.6; 5–0.8; 6–1.0

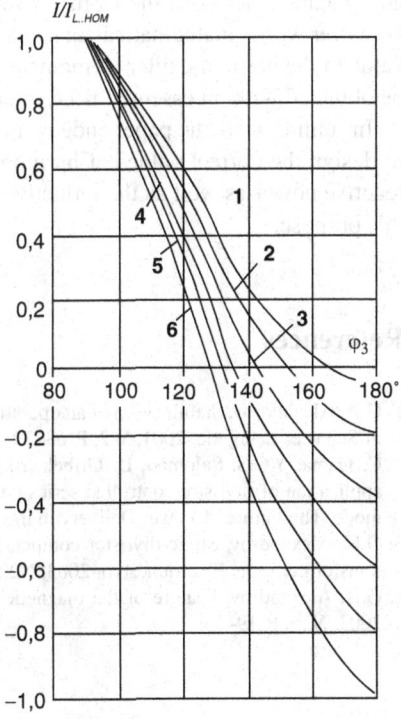

Then the relation (1.165) is substantially simplified

$$\frac{I_{(K)}}{I_{L.nom}} = \frac{(1 - K)\,\alpha}{K + \sqrt{1 + \alpha} - 1} \tag{12.27}$$

Let's define the relative value of current of thyristors complying with the passing of network winding current through zero by the curve in Fig. 12.7. For example, when $\alpha = 1$ according to the above stated $\varphi_{ig} = 129°$ and correspondingly the value I2/Imax $= 0.3$. If the equivalent reactance of thyristor limb at constant voltage is halved (K $= 0.5$), the relative value of current through the thyristors will equal to 0.3/0.5 $= 0.6$. Corresponding ignition angle of thyristors according to curve of Fig. 12.8 equals 110°. We define the relative value of CSCT current I/Inom $= 0.55$ by formula 12.27, therefore when $\alpha = 1$ that ratio corresponds with the angle $\varphi_{ig} = 110^0$. Thus all dependences from ILnom to ICnom (see Fig. 12.9) can be drawn.

12.3 Conclusion

CSCT as a new device to compensate reactive power in power systems was introduced in the paper. The main scheme of this device was also illustrated.

With a glance on the scheme of CSCT it seems to be the best way to add an additional capacitor to the ComW to have bilateral reactive power compensation. However with a mathematical analysis the idea can be changed to correct design of value of the harmonic filter elements in order to achieve a bilateral compensator can be obtained without essential need of capacitor bank.

In summarized, the paper endeavors to present final equations which are required to design the correct values of harmonic filters' elements to regulate of capacitive reactive power as well as the inductive value. The obtained results prove achieving this purpose.

References

1. G.N. Alexandrov, Stabilization of an operating voltage in electrical networks. News of Academy of Science. Energetic, 2001, W2, P. 68–79.
2. C. Gama, Y.C.S. Salomao, D. Gribek Yoao, Wo Ping. Brazilian north-south interconnection application of thyristor controlled series compensation (NCSC) to damp inter areas oscillation mode. The Future of Power Delivery in the 21st Century, 1997, Nov.
3. G.N. Alexandrov, Static thyristor compensator on the basis of a controlled shunt reactor of a transformer type. Electrichestvo, 2003, N22.
4. G.N. Alexandrov, Feature of the magnetic field of transformer under loading. Electrichestvo, 2003, N25, P. 19–26.

Chapter 13
Harmonic Analysis and Optimum Allocation of Filters in CSCT

Mohammad Golkhah and Mohammad Tavakoli Bina

Abstract A new shunt reactive power compensator, CSCT, is presented and introduced in this paper. Mathematical analysis of harmonic content of the current of CSCT is performed and use of a winding with additional circuit has been presented as a solution to suppress these harmonics.

Keywords CSCT · Harmonic filter · Reactive power compensation · Thyristor controlled transformer

13.1 Introduction

CSCT stands for "Controlled Shunt Compensator of Transformer type" [1–3]. A general scheme of this compensator is presented in Fig. 13.1. This configuration is a transformer with three windings. NW is network winding which is connected to the network and is the main winding of the compensator. CW is the second winding to which a thyristor valve and a parallel voltage circuit breaker are connected and is called CW briefly. The third winding is compensating winding which is indicated by ComW in Fig. 13.1. Two highest harmonic filters and a capacitor bank are connected to this winding. It is important to note that CSCT is a three phase compensator. The connection of NWs of three phases is star and the neutral is grounded. Control windings' connection is as same as network windings of three phases. However compensating windings can be delta in connection together.

When the thyristor is opened all the magnetic flux passes through the magnetic core leading to a minimum reluctance, maximum inductance and a capacitive current in NW according to the value of capacitor bank and eventually generate reactive power to the network. On contrary, when the thyristor is closed the flux is subjected to pass through air gap including all of the windings. Hence the reluctance, inductance and the current of NW will be maximal, minimal and maximal inductive (the rated value) respectively.

M. Golkhah (✉) and M.T. Bina
Electrical Engineering Department, K. N. Toosi University of Technology, Tehran, Iran
e-mail: goulkhah_elect@yahoo.co.uk

S.-I. Ao et al. (eds.), *Advances in Machine Learning and Data Analysis*,
Lecture Notes in Electrical Engineering 48, DOI 10.1007/978-90-481-3177-8_13,
© Springer Science+Business Media B.V. 2010

Fig. 13.1 General scheme of
a CSCT: TV-thyristors valve,
VCB-vacuum circuit breaker,
C5-L5 and C7-L7 filters of
fifth and seventh harmonics,
C-additional capacitor bank

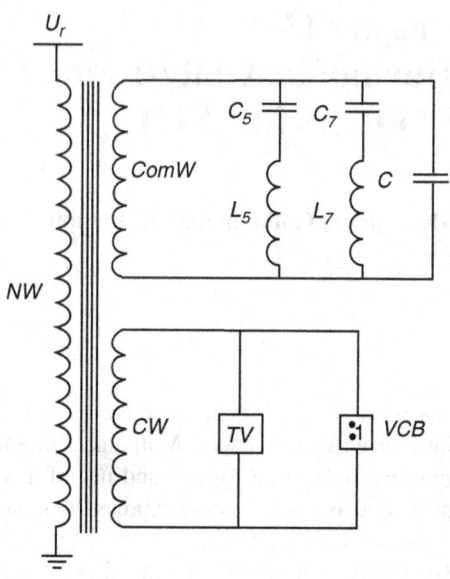

Since there is a reactive power flow control by a thyristor type switch, there will
be different order harmonics in the current of NW whose magnitudes depend on
the firing angle of the thyristor or generally the closure moment of the switch. Harmonic filters are engaged due to suppress principally the harmonics with the highest
degrees since the bigger is the degree of the harmonic, the bigger is the amplitude
of the harmonic and correspondingly the harder is elimination of that harmonic. Not
only is the harmonic filters design difficult in many cases, but also the cost of them
is usually significant. In many cases there may be different solutions to design the
filter. However there is certainly the best way to compromise between efficiency
of the filters and the costs. It is endeavored in this paper to find the best allocation and design of the harmonic filters to remove harmonic components so that both
high efficiency and low costs to be satisfied in the best way. This called for some
mathematical analysis and equivalent circuits for the windings of the transformer.

Figure 13.2 represents one phase of the transformer with mentioned three windings. The winding close to the main core is CW, the outer winding is NW and
interlayed winding is ComW.

13.2 Mathematical Analysis to Calculate Harmonic
Components of the Current

Applying thyristors to control the current of the compensator brings highest harmonics in the current. Highest harmonics are formed during incomplete combustion
angle of thyristors when current flows intermittently through thyristor block. With
angles $0 \leq \omega t \leq \psi$ and $\pi - \psi \leq \omega t \leq \pi$ the current equals zero, and with angles
$\psi < \omega t < \pi - \psi$,

Fig. 13.2 Mono phase
diagram of CSCT: (**a**) view
form side; (**b**) view form
above: 1 – a main core,
2 – yokes, 3 – lateral yokes,
4 – NW, 5 – CW, 6 – ComW,
7 – magnetic shunts

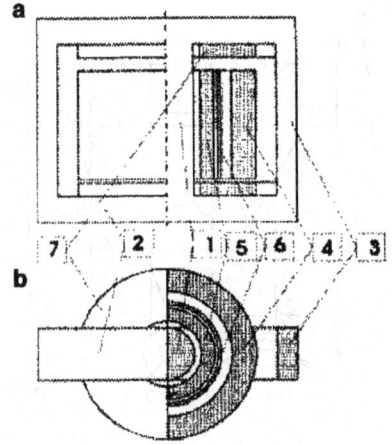

$$i(t) = I_m \cdot (\sin \omega t - \sin \psi) \tag{13.1}$$

Where the firing angle ψ can change in the range of $0 < \psi < \pi/2$.

$$I(\psi) = I_m \cdot \sqrt{\frac{1}{\pi} \int_{\psi}^{\pi-\psi} (\sin wt - \sin \psi)^2 dwt} =$$

$$= I_m \cdot \sqrt{\frac{1}{\pi} \cdot [(\pi - 2\psi) \cdot (0,5 + \sin^2 \psi) - 1,5 \sin 2\psi]} \tag{13.2}$$

of the root-mean-square current through the thyristor at arbitrary firing angle ψ to the root-mean-square of rated current $I = I_m/\sqrt{2}$ (corresponding with angle $\psi = 0$) equals to:

$$\frac{I(\psi)}{I} = \sqrt{\frac{1}{\pi} \cdot [(\pi - 2\psi) \cdot (1 + 2\sin^2 \psi) - 3\sin 2\psi]} =$$

$$= \sqrt{\left(1 - \frac{2\psi}{\pi}\right) \cdot (1 + 2\sin^2 \psi) - \frac{3}{\pi} \cdot \sin 2\psi]} \tag{13.3}$$

The current through the thyristor block decreases fast with increase in ignition angle (see Fig. 13.3). Thus the content of the higher harmonics strongly changes with change in ignition angle and can be calculated by the formula

$$k_{h.k.} = \frac{I_k}{I_1} = \frac{2}{k} \cdot \frac{\frac{\sin(k-1)\cdot\psi}{k-1} + \frac{\sin(k+1)\cdot\psi}{k+1}}{\pi - 2\psi - \sin 2\psi} \tag{13.4}$$

Where I_k and I_1 are amplitudes of k-th and fundamental harmony.

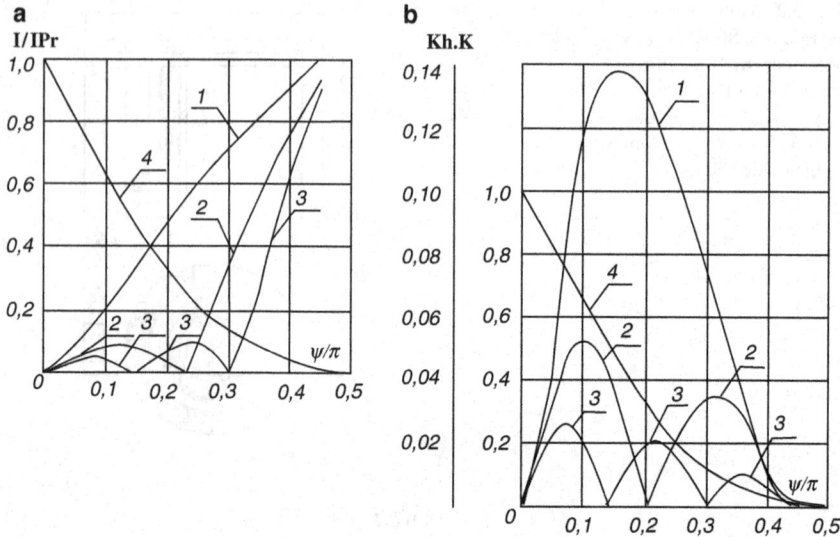

Fig. 13.3 The ratio of currents of highest harmonics: third (1). fifth (2), seventh (3) to the current of the basic frequency with dependence on ignition angle of thyristors $0 < \psi < 90°$: (**a**) in relation to present current of basic frequency; –(**b**) in relation to rated current (when thyristors are completely closed); four-ratio of root-mean-square value of full current to rated current

Results of calculations with this formula are shown in Fig. 13.3. As seen, increase in number of harmonics leads to decrease in its content. Thus the content of the third harmonic in the current continuously increases with reduction in combustion time of thyristors (increase in firing angle of thyristors). With small firing angles of thyristors (greater combustion angles), the increase in the content of the fifth harmonic is replaced by a reduction down to zero when firing angle $\psi = 0.22\pi$ and then increases again approaching 100% under a very small reactor current. The content of the seventh harmonic passes through the minimum (zero) twice and then sharply increases approaching 100% (Fig. 13.3).

The ratio of current corresponding to the highest harmonics to that of rated reactor current (to current when thyristors are completely opened) has entirely different character. For the third harmonic this attitude reaches a maximum when the reactor current is $I = 0.42 I_{nom}$ (see curves 1 and 4 of Fig. 13.3b). For the fifth harmonic the maximum is reached at the current $I = 0.63 I_{nom}$ (see curves 2 and 4 of Fig. 13.3b). For the seventh harmonic it is reached when reactor current $I = 0.72 I_{nom}$. The second maximum of the fifth harmonic is much less than the first. The second and third maximum of the seventh harmonic is also much less than the first.

Value of the first maximum in relation to amplitude of rated current, are resulted in Table 13.1 (designated by the letter β_k).

Special winding in delta connection (compensating winding) is usually used for the compensation of the highest harmonics (third). In this case for the third harmonic this compensatory winding is short-circuit, that excludes the possibility of third harmonics in magnetic flux enveloping the compensation winding.

Table 13.1 Reactive powers of capacitors and inductors related to each harmonic

k	β_k	α_k	$\dfrac{Q_k}{Q_{nom}}$	$\dfrac{Q_{c,k}}{Q_{nom}}$	$\dfrac{Q_{L,k}}{Q_{nom}}$
3	0.138	0.102	0.28	0.19	0.087
5	0.05	0.030	0.068	0.049	0.019
7	0.026	0.013	0.028	0.020	0.0073
11	0.0105	0.0044	0.0090	0.0067	0.0023
13	0.0075	0.0030	0.006	0.0045	0.0015

There are circuit designs for fifth and seventh harmonics suppression. But they are very complex and expensive. The most simple, cheap and reliable enough design is the application of filters of higher harmonics, connected to compensating windings of each of the phases. Each of such filters consists series-connected reactor (with stable inductance) and capacitors selected in such a way that they provide compensating winding short-circuit for each of the harmonics. Thus the corresponding harmonic cannot be hold in the magnetic flux engulfing compensating winding. The filter can also provide compensation of the third harmonic component of a single phase reactor.

Since the source of highest harmonics in reactor current is the control winding with thyristors, the compensating windings should cover it to exclude the possibility of keeping higher harmonics in magnetic flux and by that in the current of the network winding covering the control winding and compensating winding.

Thus for the compensation of the k-th harmonic in the current, CSCT should adhere to the equality

$$k\omega L_k = \frac{1}{k\omega C_k} \tag{13.5}$$

From there

$$\omega C_k = \frac{1}{k^2 \omega L_k} \tag{13.6}$$

Thus under no-load conditions when control block thyristors are shut (Fig. 13.4), compensating winding (ComW) of CSCT is loaded by impedance

$$X_{1,k} = \omega L_k - \frac{1}{\omega C_k} = \omega L_k \cdot (1 - k^2) = \frac{1}{\omega C_k} \frac{1 - k^2}{k^2} \tag{13.7}$$

Conformably the fundamental frequency current in control winding current of CSCT, due to filter of k-th harmonic, equals

$$I_{1,k} = \frac{U_{ph}}{\delta X_{nom} + X_{1,k}} = \frac{U_{ph}}{\delta X_{nom} + \omega L_k \cdot (1 - k^2)} = \frac{U_{ph}}{\delta X_{nom} + \frac{1}{\omega C_k} \cdot \left(\frac{1}{k^2} - 1\right)} \tag{13.8}$$

Where δX_{nom} defines the short-circuit impedance of the basic winding in relation to ComW with filters and X_{nom} also defines the rated short-circuit impedance of the

Fig. 13.4 Schematic diagram
of CSCT with higher
harmonics filters

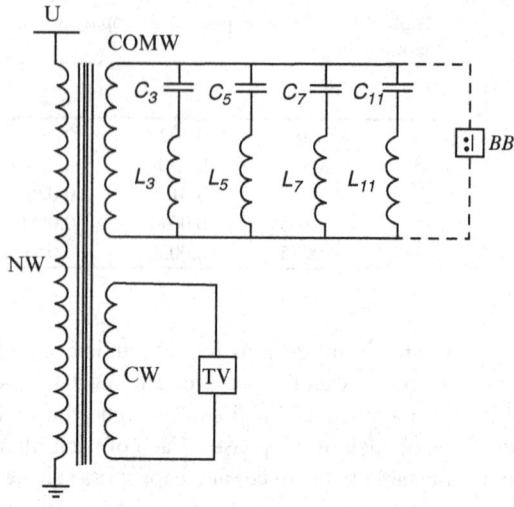

basic winding in relation to control winding (CW). Since the optimal impedance value $X1.k$ is greater than impedance δX_{nom} (see below), the current through the filter has capacitive character and the ratio of current $I1.k$ to rated current has a negative sign.

$$\frac{I_{1,k}}{I_{nom}} = \frac{X_{nom}}{\delta X_{nom} + \omega L_k \cdot (1-k^2)} = \frac{X_{nom}}{\delta X_{nom} + \frac{1}{\omega C_k} \cdot \left(\frac{1}{k^2} - 1\right)}$$

$$= -\alpha_k \approx \frac{X_{nom}}{\delta X_{nom} - \frac{1}{\omega C_k}} \qquad (13.9)$$

Where α_k is the absolute value of the ratio of basic frequency current through the k-th harmonic filter to rated reactor current.

Resolving Equation (13.9) in relation to ωL_k, we get

$$\omega L_k = X_{nom} \cdot \frac{1 + \alpha_k \delta}{\alpha_k \cdot (k^2 - 1)} \qquad (13.10)$$

Capacitor impedance of the k-th harmonic filter to power current according to (13.6), (13.10) equals to:

$$\frac{1}{\omega C_k} = X_{nom} \cdot \frac{k^2 \cdot (1 + \alpha_k \delta)}{\alpha_k \cdot (k^2 - 1)} \qquad (13.11)$$

Power of the k-th harmonic filter chokes, due to basic frequency current is:

$$Q_{Lk} = I_{1.k}^2 \cdot \omega \cdot L_k = \alpha_k^2 \cdot I_{nom}^2 \cdot X_{nom} \cdot \frac{1 + \alpha_k \delta}{\alpha_k \cdot (k^2 - 1)} \qquad (13.12)$$

Capacitor power of the same filter is:

$$Q_{C_k} = I_{1.k}^2 \cdot \frac{1}{\omega \cdot C_k} = \alpha_k^2 \cdot I_{nom}^2 \cdot X_{nom} \cdot \frac{k^2(1 + \alpha_k \delta)}{\alpha_k \cdot (k^2 - 1)} \qquad (13.13)$$

Total absolute k-th harmonic filter power due to basic frequency current

$$Q_{\Sigma.k} = Q_{L_k} + Q_{C_k} = \alpha_k^2 \cdot Q_{nom} \cdot \frac{(1 + k^2)(1 + \alpha_k \delta)}{k^2 - 1} \qquad (13.14)$$

Where the rated power of one phase of the reactor equals

$$Q_{nom} = I_{nom}^2 \cdot X_{nom} \qquad (13.15)$$

The maximum current of the k-th harmonic through the filter of that harmonic can be calculated analytically and defined according to (13.4) by

$$I_k^= \beta_k \cdot I_{nom} \qquad (13.16)$$

Accordingly total absolute power of k-th harmonic filter, due to current of k-th harmonic

$$Q_{k.k.} = I_k^2 \cdot \left(k \cdot \omega \cdot L_k + \frac{1}{k \cdot \omega \cdot C_k} \right) = 2I_k^2 \cdot k \cdot \omega \cdot L_k =$$

$$= 2\beta_k^2 \cdot I_{nom}^2 \cdot X_{nom} \frac{k \cdot (1 + \alpha_k \cdot \delta)}{\alpha_k \cdot (k^2 - 1)} = 2\beta_k^2 \cdot Q_{nom} \frac{k \cdot (1 + \alpha_k \cdot \delta)}{\alpha_k \cdot (k^2 - 1)} \qquad (13.17)$$

Total absolute power of k-th harmonic filter, due to currents of basic and k-th harmonics equals to:

$$Q_k = Q_{k.1} + Q_{k.k} = Q_{nom} \cdot \frac{1 + \alpha_k \cdot \delta}{k^2 - 1} \cdot \left[\alpha_k \cdot (1 + k^2) + 2\beta_k^2 \cdot \frac{k}{\alpha_k} \right] \qquad (13.18)$$

We shall find the optimum value Q_k by equating zero derivative of Q_k on α_k

$$\frac{\partial Q_k}{\partial \alpha_k} = \frac{1}{k^2 - 1} \cdot \left[1 + k^2 - 2 \cdot \frac{\beta_k^2 \cdot k}{\alpha_k^2} + 2\alpha_k \cdot \delta \cdot (1 + k^2) \right] \cdot Q_{nom} = 0 \qquad (13.19)$$

From last equation we get the value α_k corresponding to the minimum power of k-th harmonic filter

$$\alpha_k = \beta_k \cdot \sqrt{\frac{2k}{(1 + k^2) \cdot (1 + 2\alpha_k \cdot \delta)}} \qquad (13.20)$$

In this solution, α_k also contains a small term under a root. The smallness of this term allows to calculate α_k by method of successive approximation, assuming as a first approximation that $\alpha_k = 0$ or $\alpha_k = \beta_k$. Taking calculation data for highest harmonic according to data of the following table and estimating the value $\delta = 0.5$, we obtain the following values of α_k and corresponding filter capacity according to (13.17), and also relative values of power capacitors $(Q_{c.k})$ and reactors $(Q_{L.k})$ of filters subject to high-frequency current component (see Table 13.1).

It follows from the resulted data, that the capacity of filters contains a small part of CSCT capacity, especially in the case when the third harmonic is compensated by compensating windings in delta connection of the three phases of CSCT. In this case total capacity of filters does not exceed 10% of reactor power.

To estimate the efficiency of highest harmonics restrictions in reactor current, we shall consider its equivalent circuit in resonance mode in the k-th harmonics (Fig. 13.5). We shall estimate parameters of CSCT equivalent circuit according to Fig. 13.4. In this case the equivalent cross-section of magnetic flux linked with network winding (NW), during short-circuit of control winding (CW) is:

$$F_{eff.1} \approx \pi \cdot d_{12} \cdot \left(a_{12} + \frac{a_1 + a_2}{3} \right) \tag{13.21}$$

Where d12 – mean gap diameter between CW and NW, a12 – gap thickness (radial size), a1 and a2 – thicknesses of CW and NW (radial size).

Equivalent cross-section of magnetic flux linked with CW, during short-circuited ComW, in which filters are connected in parallel,

$$F_{eff.2} \approx \pi \cdot d_{13} \cdot \left(a_{13} + \frac{a_1 + a_3}{3} \right) \tag{13.22}$$

Where d13 – mean gap diameter between ComW and CW, a3 – thickness (radial size) of ComW $(a_3 \approx 0.3 a_2)$.

If ComW is located in the middle of CW and NW, $d_{13} = d_{12} + a_{13}; a_{13} = 0.5 a_{12}$

Equivalent cross-section of magnetic flux linked with ComW during short-circuited CW,

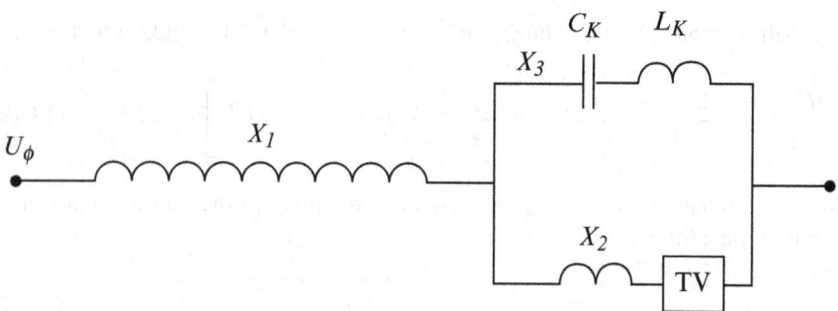

Fig. 13.5 Equivalent three-beam scheme of CSCT

$$F_{eff.3} \approx \pi \cdot d_{23} \cdot \left(a_{23} + \frac{a_2 + a_3}{3} \right) \qquad (13.23)$$

Where d_{23} – mean diameter of a gap between CW and ComW, a_{23} – thickness of that gap.

In the particular case, where ComW is positioned in the middle of CW and NW, considering that the thickness of ComW is small in comparison with CW and NW, we get $a_{23} = 0.5a_{12}$, $d_{23} = d_{12} - a_{23}$.

Correspondingly, the short-circuit impedance of NW in relation to CW equals

$$X_{12} = \frac{8 \cdot 10^{-7} \cdot \pi^2 \cdot f \cdot N_1^2 \cdot F_{eff.1}}{l_0} = X_{min} \qquad (13.24)$$

Where $N1$ – number of turns of NW, l_0 is height of magnetic conductor window.

Short-circuit impedance of NW in relation to ComW of closed filters,

$$X_{13} = \frac{8 \cdot 10^{-7} \cdot \pi^2 \cdot f \cdot N_1^2 \cdot F_{eff.2}}{l_0} = X_{min} \cdot \frac{F_{eff.2}}{F_{eff.1}} = \delta \cdot X_{min} \qquad (13.25)$$

For example, it is always possible to choose the position of ComW so that $\delta = 0.5$.

Short-circuit impedance of ComW in relation to CW

$$X_{23} = \frac{8 \cdot 10^{-7} \cdot \pi^2 \cdot f \cdot N_1^2 \cdot F_{eff.3}}{l_0} = (1 - \delta) \cdot X_{min} \qquad (13.26)$$

When $\delta = 0.5$, $X_{23} = 0.5 X_{min}$.

The parameters of k-th harmonics equivalent three-beam scheme of CSCT from the deduced relations equals (see Fig. 13.5).

$$\left. \begin{array}{l} X_{1.k} = \frac{k}{2} \cdot (X_{12} + X_{13} - X_{23}) = k\delta \cdot X_{min}; \\ X_{2.k} = \frac{k}{2} \cdot (X_{12} + X_{23} - X_{13}) = k \cdot (1 - \delta) \cdot X_{min}; \\ X_{3.k} = \frac{k}{2} \cdot (X_{13} + X_{23} - X_{12}) = 0. \end{array} \right\} \qquad (13.27)$$

Hence, the equivalent circuit for k-th harmonics looks like represented in Fig. 13.6, where the thyristor block is equivalent to the current generator. Apparently, in this case all the current of k-th harmonic become locked in the filter and does not get in the network winding.

ComW suppresses higher harmonics most effectively when it is positioned in between CW and NW.

Further it is necessary to find out the influence of the presence of highest harmonics filters on CSCT rated current. Nominal condition complies with the complete closing of thyristors when highest harmonics in reactor current are absent.

The equivalent circuit for the first harmonics when the thyristors are completely closed is represented in Fig. 13.7. Equivalent impedance of branch 3 with filter equals to the impedance of the filter (13.7)

Fig. 13.6 Equivalent
three-beam scheme of CSCT
for k-th harmonic when
ComW is positioned between
CW and NW

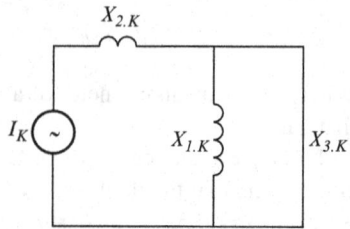

Fig. 13.7 Equivalent circuit
of CSCT in normal mode for
calculating power current
considering k-th harmonic
filter

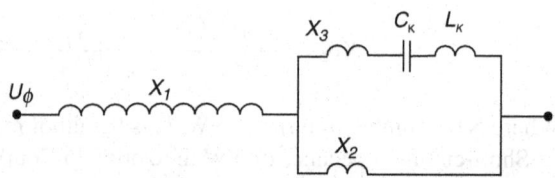

$$X_{3.eq} \cdot \omega \cdot L_k \cdot (1 - k^2) =$$
$$= X_{min} \cdot \frac{1 + \delta \cdot \alpha_k}{(k^2 - 1) \cdot \alpha_k} \cdot (1 - k^2) = -X_{min} \cdot \frac{1 + \delta \cdot \alpha_k}{\alpha_k} \qquad (13.28)$$

Equivalent impedance of branch 2 in accordance with (13.25) equals to $X_{2.eq} = (1 - \delta) \cdot X_{min}$.

Equivalent impedance for the parallel connection of branches 2 and 3

$$X_{2.3.eq} = \frac{X_{2.eq} \cdot X_{3.eq}}{X_{2.eq} + X_{3.eq}} = X_{min} \cdot \frac{1 - \delta}{1 - \frac{\alpha_k \cdot (1 - \delta)}{1 + \delta \cdot \alpha_k}} \qquad (13.29)$$

Total equivalent impedance of CSCT in normal mode considering k-th harmonic filter in accordance with (13.25), (13.27) equals

$$X_{eq.\,min} = X_1 + X_{2.3.eq} = \delta \cdot X_{min} + X_{min} \frac{1 - \delta}{1 - \frac{\alpha_k \cdot (1 - \delta)}{1 + \delta \cdot \alpha_k}} \qquad (13.30)$$
$$= X_{min} \frac{1 + \delta^2 \alpha_k}{[1 + \alpha_k (2\delta - 1)]}$$

For example when $\delta = 0.5$ considering the third harmonic filter with the biggest current (see Table 13.1, $\alpha_3 = 0.102$), we shall get

$$X_{eq.\,min} = X_{min} \cdot \frac{1 + 0.102 \cdot 0.5^2}{(1 + 0.102 \cdot 0)} = 1.025 \cdot X_{min} \qquad (13.31)$$

In that case, the presence of filters greatly reduces the rated current of CSCT (approximately by 3%) in comparison with CSCT without filters.

It is necessary to note that the presence of compensation winding with filters between CW and NW allows the provision of short-term forced capacity of the reactor which is necessary for drastic restriction of switching overvoltages. In as much as the inductive impedance of CSCT at short circuit ComW is less than at short circuit CW ($\delta < 1$), short-term short-circuit of ComW, for example, with vacuum switch (VCB on Fig. 13.4), will lead to forced capacity of CSCT about $1/\delta$ times. When $\delta = 0.5$ the forced reactor capacity will exceed nominal two times.

The cross-section of compensation winding (ComW) accordingly as stated above should be chosen the course of all currents. During delta connection of ComW it is necessary to consider the current of the third harmonic in the triangle according to Table 13.1.

$$I_3 = 0.138 I_{1.\max} \cdot k_{tr}$$

Where k_{tr} is ratio of turns in NW and ComW.

Power current and fifth harmonic current through fifth harmonic filter

$$I_5 = (0.03 + 0.05) \cdot I_{1.\max} \cdot k_{tr}$$

Power current and seventh harmonic current through seventh harmonic filter

$$I_7 = (0.013 + 0.025) \cdot I_{1.\max} \cdot k_{tr}$$

As such, total current through ComW equals

$$I_{Com.\Sigma} = 0.256 \cdot I_{1.\max} \cdot k_{tr}$$

Considering the discrepancy of the maximum of all currents components, it is possible to reduce the rated current of ComW, in delta connection and take

$$I_{Com.nom} = 0.2 I_{1.\max} \cdot k_{tr}$$

It is necessary to consider presence of the third harmonic filter when the triangle of ComW is opened (by its star connection) and respectively in addition to consider the power current of the third harmonics filter. As a result, the total current of ComW in this case equals

$$I_{Com.\Sigma} = 0.358 I_{1.\max} \cdot k_{tr}$$

Considering the discrepancy of the maximum of all currents components, it is possible to reduce the rated current of ComW and take

$$I_{Com.nom} = 0.3 I_{1.\max} \cdot k_{tr}$$

The cross-section of conductor ComW equals

$$F_{co.Com} = \frac{I_{Com.nom}}{J_{Com.opt}}$$

The volume of copper of ComW equals

$$V_{co.Com} = \pi \cdot d_{Com.av} \cdot N_{Com} \cdot F_{co.Com}$$

Where NCom $= \frac{N_1}{k_{tr}}$.

It is necessary to add this volume of copper to the total volume of copper of operated reactors. For example, for reactor with ComW in star connection:

$$V_{co.Com} = \pi \cdot d_{Com.av} \cdot \frac{N_1}{k_{tr}} \cdot \frac{0.3 \cdot I_{1.max} k_{tr}}{J_{Com.opt}} = \pi \cdot d_{Com.av} \cdot \frac{0.3 N_1 \cdot I_{1.max}}{J_{Com.opt}} \quad (13.32)$$

13.3 Conclusion

CSCT as a new device to compensate reactive power in power systems was introduced in the paper. The main scheme of this device was also presented and illustrated.

Using a thyristor to control the current of the compensator leads to appearance of harmonics in the current. Basic equations of highest harmonics in the current of the compensator are presented. Moreover adding a third winding with highest harmonic filters in less voltage and more current levels than NW is presented as a solution to suppress the harmonics. It is demonstrated in this paper that the optimum place to emplace this winding is between NW and CW.

Obtained results prove that the content of harmonics in the current of CSCT is less than 2% and this fact denotes a successful design of damping surplus harmonics in the current of CSCT.

References

1. G.N. Alexandrov, Stabilization of an operating voltage in electrical networks. News of Academy of Science. Energetic, 2001, W2, 68–79.
2. C. Gama, Y.C.S. Salomao, D. Gribek Yoao, Wo Ping. Brazilian north–south interconnection application of thyristor controlled series compensation (NCSC) to damp inter areas oscillation mode. The Future of Power Delivery in the 21st Century, Nov. 1997.
3. G.N. Alexandrov, Feature of the magnetic field of transformer under loading. Electrichestvo, 2003, N25, 19–26.

Chapter 14
Digital Pen and Paper Technology as a Means of Classroom Administration Relief
Mobile Tools for Teachers (MTT) Evaluation Study

Jan Broer, Tim Wendisch, and Nina Willms

Abstract This paper contains the results of the Mobile Tools for Teachers project concerning the viability of digital pen and paper technology (DPPT) for administration in a K-12 classroom environment. Filled out forms were evaluated and interviews as well as user tests with teachers were done to show the advantages and disadvantages of DPPT compared to regular methods for attendance tracking and grading. Additionally, the paper addresses the problems that arise with DPPT in a classroom environment and includes suggestions how to deal with those.

Keywords Digital pen and paper technology · Classroom administration · Attendance tracking · Mobile tools for teachers

14.1 Introduction

The German school system, like many other school systems all over the world, has been reformed in the past years to allow for a better documentation and evaluation of teaching processes. Administrative tasks are an integral part of teachers' workloads these days and they have to fulfill many roles aside from teaching itself. This shifts workload into teachers' free-time and reduces the time they have for regeneration. An OECD study on the topic reports that recent reforms will have a strong influence on mechanics that traditionally steer the work of teachers [7] and that there's generally very little support for schools in the field of management and administration [7]. Upcoming reforms are expected to require even more data collection in the area of classroom management. Johannes Hartig describes the goal of these reforms as making the productivity of educational systems, the quality of individual educational institutions, and the learning success of individual students measurable [8].

Classroom administration is becoming more and more of an issue for K-12 teachers. A growing requirement to document the teaching processes and student

J. Broer (✉), T. Wendisch, and N. Willms
University of Bremen
e-mail: scrusi@gmail.com; tim.wendisch@s-hb.de; ninawillms@gmx.net

S.-I. Ao et al. (eds.), *Advances in Machine Learning and Data Analysis*,
Lecture Notes in Electrical Engineering 48, DOI 10.1007/978-90-481-3177-8_14,
© Springer Science+Business Media B.V. 2010

developments leads to teachers using more and more time for administrative tasks and them having less time to focus on individual students. School avoidance is often detected too late and it is hard for teachers to give individual feedback to students or their parents if they teach large classes and have to spend more and more time on documentation tasks [3].

The standard German system of tracking attendance information with a class book is cumbersome and highly insecure as no backup is created and data have to be tallied up by hand at the end of a semester, Alarming trends are often only realized at that point as individual teachers rarely communicate their issues with students to one another or the principal. See Refs. [1, 3] for a detailed explanation of these problems.

The 1 year long master project Mobile Tools for Teachers (MTT) at the University of Bremen implemented and evaluated a prototypical system that uses digital pen and paper technology (DPPT) for classroom administration. DPPT uses a special pen and paper with a special pattern printed on it to capture handwritten information. The system allows teachers to track attendance and grading information on paper while digitalizing said information at the same time. This way a copy of all information is created as well as a central point where authorized persons can view the data. The assumed advantages of the system were twofold: On the one hand the system was designed to save time spent on classroom administration by eliminating such tasks as sifting through handwritten attendance information at the end of a semester. On the other hand the project aimed at giving various parties better access to information that is collected in schools, allowing for example for early detection of school avoidance and faster finding of final grades. This paper deals with the evaluation of how well these goals were achieved and how suited DPPT in this setup is for classroom environments in general.

We assumed that a system using pen and paper would require next to no training on the side of the users and would be much more likely to be accepted by users who are not very familiar with computer technology. This assumption was based on informal interviews done with a few teachers at the start of the project as well as findings other scientists had with different technologies in a teaching environment. Jocelyn Wishart, Angela McFarlane and Andy Ramsden from the University of Bristol, UK for example, studied the usage of personal digital assistants (PDAs) among science teachers in training. They found out that the "use of their PDAs dropped during the main 12 week block of teaching practice when they are placed in schools full-time. Some trainees reported that under pressure of time and workload they reverted to use of paper and pen to organise themselves and plan their teaching" [12].

Additionally it appears, that graphical interfaces can distract from the actual task at hand. Sharon Oviatt, Alex Arthur, and Julia Cohen compared math students' use of different technologies for solving math problems. They found that students' "ability to think at a more abstract, strategic, and self-reflective level about the process of solving math problems declined significantly when using the graphical tablet interface" [10].

If computer based systems such as PDAs are too complicated to use in stressful situations, it seems natural to use a tool that allows teachers to stick to pen and paper while still providing the functionalities of a digital system. Various models on the acceptance of technology, such as the Technology Acceptance Model TAM [4] and the Structural Model of Educational Technology Acceptance Model Integrated with Educational Ideologies. The MTT system was implemented with the cooperation of and tested at the International School of Bremen (ISB), a "private, co-educational, college-preparatory, English-speaking school" (www.isbremen.de). Classes in the ISB are generally smaller than those in public German schools. We originally intended to work with the German Wilhelm-Olbers-School[1] as well, however cooperation was slow and had to be terminated after a few meetings. Our time limits in combination with school response times did not allow us to switch to another school.

14.1.1   Attendance Tracking and Grading in German Schools

Whereas the MTT system was eventually tailored to the needs of the teachers at the ISB, its core elements were designed with the standard German system in mind. We used a combination of our own experiences in German schools and the investigation at the Wilhelm-Olbers-School in Bremen as a basis for this. German schools usually use class books or course books to track attendance data and general lesson information [9]. Normally one book is used for each class and for each semester, some schools split students into courses instead, making one book per course and semester necessary. These books contain two types of information – general information about the class and information about individual lessons.

The general part of such a book consists mostly of a list of all students who are in that class or course, a general timetable, and a list of exams which were written or will be written. The daily part consists of detailed information about the content covered that day, as well as students' attendance. The main structural element of each page is the day's timetable. In this timetable there are columns for subject, content covered in that lesson, attendance and the teacher's signature. Some of the information entered on such a sheet has to be entered each day, some is optional. Class books are often handled by a designated student in each class who carries them around to the different teachers. Course books on the other hand often stay with the teacher. Grade tracking is usually not standardized within a school. It is up to each individual teacher how they grade their students and how they keep track of these grades. The calculation of a final grade from intermediate grades is also up to the teachers, leaving them to find their own system of weighting different grading types. Commonly used methods include small grading books with rows for each student and spreadsheet software.

---

[1] http://www.szdrebberstrasse.de/.

14.1.2 Special Case: The International School of Bremen

The system at the ISB is different to what German schools use, much more akin to the system in the USA. Teachers have fixed rooms and the students always have to switch rooms for the next class, contrary to traditional German schools where most of the lessons are given in the class's classroom [9]. Attendance is only checked once per day at the ISB, from 8 to 8:10 am. The students have to meet with their class teacher in his or her room at that time so that attendance can be taken.

When we started working with the ISB in 2007, each teacher had a list of his students on a sheet of paper. The teacher filled out this list in the morning and sent a student downstairs to the secretary who checked whether a missing student was excused or not and took appropriate action. Normally this process took the whole 10 min allocated to it, from the beginning of attendance tracking until the student returned.

During our collaboration with the ISB, they switched to the computer based management information system Facility by the Serco Group plc.[2] Due to the existing infrastructure at the school – each teacher has a PC in his or her room – it is possible for them to track attendance using a computer. The process now is as follows:

The teacher has to log into the Facility system with his or her personal login and password. Then he or she selects the class list, which has two columns. The first for the name of the students (already inserted) and the second is for a character from the legend (such as E for excused or L for late). The legend consists of nearly 20 possible characters but only 3 to 4 were used regularly. After filling out this list, the teachers save it and the secretaries have to check all lists on an overview page on their own computer.

Ideally the whole process takes about 2.5 min, but it can be prolonged by the class (number of students, discussions, questions, late comers, etc.), by the computer (speed) and by the teacher himself. Grading at the ISB is done similar to German schools – in a non-standardized way. A seemingly cosmetic difference is that the ISB uses grades from 7 to 1, where 7 is the best grade.[3]

14.1.3 Digital Pen and Paper Technology

The digital pen and paper technology referred to in this document was developed by the Lund, Sweden based Anoto Group AB.[4] A ball point pen is combined with a digital camera and a storage and processing unit to form the digital pen. Writing on paper that was pre-printed with a special pattern of small black dots allows the camera to obtain position information. This pattern is usually found on forms that have been designed for the use with the digital pen, consisting of text fields (free

---

[2] http://www.serco.com.

[3] As opposed tot he traditional German grades from 1 to 6 where 1 is the best grade.

[4] http://www.anoto.com.

and dictionary based), check boxes and special command boxes called pidgets. All data collected with the pen are then mapped to an electronic representation of the form layout, providing meaningful information.

This DPPT implementation makes a difference between so-called unique pattern and copied pattern. If a page is printed with unique pattern that means that the pen can identify the exact page you are writing on through the alignment of the dots. The less expensive approach is the copied pattern one, which uses the same pattern on each sheet of the same type. This saves licensing and printing costs but makes it impossible to uniquely identify the sheet that was written on, usually solved through user interaction (i.e. filling out a date or ID field).

The MTT project used a server provided by the Bremen, Germany based bendit GmbH[5] for this mapping and the subsequent character recognition. The bendit software combines stroke information from the pen with a dictionary approach to obtain better recognition rates.

14.1.4 The MTT System

The MTT project implemented a prototype of a system that uses digital pens developed by Anoto to track attendance and grading information written on special sheets of paper, the MTTForms. These pens transfer their data via USB to a personal computer that then sends them via the internet to a server for processing. After intelligent character recognition[6] is performed, all data are stored in a database and can subsequently be accessed through a web interface called MTTWeb. The web system allows for different views on the collected information, such as graphs showing the development of student grades over time.

The attendance tracking process was modeled closely on the principle of the German class book, using forms that should be intuitive to use for anyone that has used a class book before. Grades are tracked on special grading sheets which are pre-filled with student names and allow teachers to quickly write down grades for different types of examinations. A detailed description of the MTT system can be found in Ref. [1].

14.1.5 Previous Studies on the Digital Pen and Paper Technology

The DPPT has been used in a number of pilot projects previously. Scientists at the University Hospitals of Geneva, Switzerland tested the technology in a clinical

---

[5] http://www.digipen.de.

[6] The term intelligent character recognition (ICR) is used here instead of optical character recognition because it relies on stroke information and not pixel information and because it is able to use dictionaries for a better recognition rate. The ICR used by the bendit GmbH is not a learning system, however.

environment [5]. While they collected positive user feedback in general, they identified some issues with both handling of the pen and data accuracy. Users in their study wrote 30% of all data entered outside the predefined boxes, making them invisible to the system. The group did not get any results on the quality of intelligent character recognition, but stated that "the quality of data obtained with the digital pen was always less or equal to that obtained with a scanner, when performed without any additional human intervention" [5]. They did not compare the results of the pen to those acquired manually.

Similar results come to us from a group of scientists at various hospitals in Bonn, Germany [6]. They used the pen to capture vital sign data in acute care settings and concluded "that data do need to be verified before they are transferred to the repository" [6]. They also found a high general acceptance of the technology among study participants.

14.2   Methodology

The study described here had the goal to prove or disprove the following assumptions: DPPT as used in the MTT system saves time, DPPT only requires minimal amounts of training and can easily replace previous methods and tools for attendance tracking and grading, and average users will have more problems with the web interface than with the pen.

The evaluation of a concept with large scale and long term results such as the MTT project is inherently difficult in a limited time frame. After completing the prototypical system, the project members had 3 weeks time to test it on actual teachers. Ideally one would have a large group of teachers with different backgrounds use the whole system for at least one semester to see how it performs in an actual use context. In order to still get relevant results we collected attendance and grading data long before the system was actually in place, so that the web interface could show real data for the teachers to use. These data were manually entered into the database, Still the data in the system were not complete due to several issues such as a low form return rate at the start and incorrectly filled out forms, and could therefore not be used to perform the teachers' actual tasks. To circumvent this problem we designed a number of problems that the teachers were asked to solve under supervision. Six of initially eight teachers at the ISB participated in the experiment. These teachers had been working with the MTT forms for a few months and were trained by project members in their use. They also got a brief introduction into the web system before they started the experiment.

14.2.1   Practical Test and Observation

Each teacher was handed a short instruction booklet on how to use the system, a list of five tasks to perform with the digital pen and in MTTWeb, a grading sheet, a class book sheet and got access to the web system. Using the digital pen they had been handed previously, they performed the following two tasks:

1. Imagine you performed a test in one of your classes last week. All students except for [name removed] (tenth grade) or [name removed] (ninth grade) respectively were present. The absent student was excused. One of the students (choose one) was present but handed in an empty sheet of paper for the test.

   - Give grades as you see fit
   - Synchronize the pen with your computer

2. You are the first teacher to fill out the class book sheet that day.

   - Fill out the form for the situation described in task 1

All participating teachers had been using the digital pens and the MTTForms for several weeks and were therefore expected to have little problems with these two tasks. The following three tasks incorporated the web system which the teachers had only seen once previously in a 15 min presentation.

1. [name removed] (ninth grade) or [name removed] (tenth grade) came to you after the test was returned and pointed out that he or she should have gotten a better grade. You agree. Log into MTTWeb and change the grade in question.
2. You need to come up with a final grade [name removed] (ninth grade) or [name removed] (tenth grade) respectively. Use the grading aid function to determine a final grade.
3. You are preparing for a meeting with the parents of [name removed] (ninth grade) or [name removed] (tenth grade) respectively and want to use MTTWeb to collect information on that student beforehand. Collect the following data by either printing or saving them in a pc document (i.e. using MS Word).

   - Exam and oral grades of the student in your class throughout the semester
   - Attendance of the student in your class during the semester
   - Average grade development of students in your class throughout the semester

These tasks were carried out on each teacher's personal computers in the school. Each teacher was observed by a project member throughout the process using specially created observation sheets, modeled on the qualitative research methods laid out by Bogdan and Taylor [2]. The observation sheet for the first two tasks asked for the time taken to perform the tasks, detailed competence in using the forms and the pens as well as a general overview on how well the individual teachers did. The three tasks concerning the frontend MTTWeb had their own observation sheet, looking again for the time taken, competence and problems using the system and general observation. Both sheets included room for handwritten comments by the observer.

Most parts of the observation sheets where designed to be easily quantifiable and could therefore be entered into and statistically evaluated with spreadsheets. The individual comments were condensed in a text document, clustered and then statistically evaluated.

14.2.2 Interviews

Each of the teachers that did the tasks described above was afterwards interviewed to get a more in depth view on how they performed using the system. These interviews were done by six different interviewers and therefore potentially biased. In order to minimize the influence of the interviewers on the results, they were handed detailed interview guidelines. These contained a combination of questions with binary (yes/no) or numerical (seven point scale) answers and open free text questions. The interviews focused on the tasks described above and were designed to answer the questions of this evaluation.

Binary and numerical answers could easily be quantified and used for statistical analysis, the free text answers where condensed in a text document, repeated answers identified and categorized as positive or negative factors for the system.

14.2.3 Form Analysis

In addition to the direct experiments with the teachers of the ISB, we also collected all the forms they filled out within the project timeframe as well as the data transferred to our web server as soon as the teachers were equipped with the digital pens. A set of guidelines we developed allowed the categorization of mistakes on the forms and problems with the intelligent character recognition. With these tools we could then quantify issues on the form side of the system.

14.3 Results and Discussion

This chapter answers the questions brought up in the introduction, split up into a general evaluation of a DPPT system in schools (using the ISB as an example) and the specific evaluation of the MTT System.

14.3.1 Viability of the Digital Pen in a Classroom Environment

One of the main assumptions that lead to the MTT system was that the digital pen is a highly viable tool for the use in a classroom environment. The high mobility and low learning threshold of a system that is based on pen and paper are the major distinctions from implementations that use technology such as the PDA or personal computers. Five of the six interviewees told us that they could imagine using the digital pen every day and none said that they could not complete the tasks they were given for data entry. Multiple teachers also commented that the digital pen was easy to use and felt good in their hand. Only one found it too bulky and not actually a handy tool.

This high acceptance was not reflected very well in the return of forms, both on paper and through the digital system. Only two of the teachers regularly used the digital pen to add data to the web system throughout the 3 week testing period. This, however, is at least partially due to the lacking test setup with only eight teachers that had to use our system in parallel to their normal one. It can safely be assumed that the return would be much higher if the system was implemented for a complete school as the only system, therefore not adding additional workload to the teachers' schedule. Three teachers saw as a problem that pens could not be kept safe in schools and four admitted problems with using the same pen each time. The first problem is a limitation of every mobile technology – the higher the mobility of a piece of hardware the easier it is to steal or lose. Stealing should be less of a problem with the digital pens though, as they currently can only be used with the system they were licensed for. This fact probably won't change unless the technology is made available as stand-alone software. Using the same pen each time to track attendance and grades is more likely to be a problem. Most teachers seem to manage to keep their personal notebooks safe and with them, however, and appropriate training should enable them to do the same with the digital pen.

The pen is, as mentioned by one of the teachers, an additional device that needs to be kept around and safe and teachers have the additional workload of connecting it with their computer. Observation showed little problems in the handling of the digital pen, with only one teacher using the wrong pen for their tasks and only one observer reporting user problems with the synchronization.

14.3.2 Viability of Static Forms in a Classroom Environment

Multiple problems appeared with the use of the MTTForms that are inherently connected to the DPPT and likely not caused by our form design. One absolutely vital element of the copied pattern approach we used for the implementation is the send pidget. The user has to hit one of these special form elements with the pen after he or she completes filling a form. If they don't, the form is not transmitted when the pen is in the cradle and the entries of multiple forms may be overlaid on top of each other, rendering all data collected unusable. This problem arose on 47.2% of the 36 grading forms we collected after the system was officially introduced at the ISB and in 51.1% of the 90 occasions where the pidget should have been checked on the class book sheets. This indicates that the teachers were not able to use the forms correctly without understanding the underlying technology. DPPT in the setup we used is restrictive in many areas beyond the pidget issue. Unlike a human evaluator, the intelligent character recognition software can only match entered data to a certain subject if the right area for entry is used. Writing outside of text boxes or across lines are unsolvable problems for the software as it is. Computational evaluation of handwriting also requires that the users stick to certain conventions and failing to fill out a certain field or not filling it out correctly can quickly lead to a whole sheet of data being lost. Further analysis of those 36 grading forms showed more errors that seem based on not understanding the technology: Empty comment fields filled with a dash (9.6%), text corrected directly on the page and not using the correction function (7.3%), text written across the border of input fields (4.5%), and data entered that cannot be understood by the MTT system (6.6%). This did not significantly improve through training.

Training did improve those parts of the form usage that were understandable for the teachers, however. In the beginning of the form usage, absent students were marked incorrectly 48.5% (n = 65) of the time. This reduced to 8.9% (n = 90) in the final phase of testing. Many of these problems can be avoided by using a unique pattern approach, but this is so expensive in licensing and printing costs that it cannot be used in a large-scale implementation so far.

14.3.3 Viability of Text Recognition

The digital pen functions very well when the forms used consist mostly out of check-boxes, recognition here is close to 100%. Text recognition is more of a problem and can be subdivided into two categories for the system provided to us by the bendit GmbH. Text fields can either be based on a white list of words, basically limiting recognition to a dictionary tailored to the application, or on plain character recognition. The first approach is useful for fields that only have a limited amount of words that can be entered into them, such as the subject field in a class book. Technical word recognition rates[7] in these fields approached 90% throughout our tests. Often a figure of 90% would be seen as very low for a character recognition system, and it is in this case as well – considering that it means that every tenth entry into a class book will be unusable this way. However, compared to the free text recognition this is a very high recognition rate. Our six results show 62% recognition of words in free text fields and only 45% of numbers in those fields were recognized correctly.[8] These problems are not only based on the intelligent character recognition system we used, but also on the previously mentioned issues of writing outside textboxes or across lines. Still, neither recognition rate is acceptable for a large-scale implementation, rendering the use of DPPT as a whole ineffective unless improved.

14.3.4 Benefits of the MTT System in a Classroom Environment

The MTT System was designed as a tool that saves teachers time and allows them to have a better overview over what is going on in their classes. Especially schools with

---

[7] A word was counted as incorrect if it was not recognized exactly as it was written on the paper. All words that were still understandable by humans but not technically correct were counted as incorrect.

[8] This evaluation was done on the basis of 100 randomly chosen text fields on the submitted forms. This evaluation was done on the basis of 100 randomly chosen text fields on the submitted forms.

large classes would profit from the system as the administrative effort rises with the amount of students in a class. The results obtained in the ISB may therefore not be generalized to have a meaning for normal German schools. It can safely be assumed that the benefits of our system are higher in those schools than what we observed in the ISB because teachers cannot stay in contact with a large number of students as well as they can with a small number.

Cleaned time measurements[9] show that teachers took only slightly longer to take attendance with the MTT system (2.8 min on average) than they did with their normal system (2.5 min on average). This allows the conclusion that the MTT system with fully trained users would be about as fast in attendance tracking as the PC based system the ISB currently uses. Teachers mentioned to us, however, that it takes them a certain time to start up the computer which is obviously not needed for the pen. If the machines used for the PC based attendance tracking stay on all day that is not a problem, but if teachers need to start up the machines every time before they can track attendance, the MTT system gets a strong time advantage.

Further hints at a time advantage were found by other members of the MTT project. Rahamatullah and Trappe performed a survey in various international schools and found out that teachers need, on average, 1 h and 47 min to transfer their grading data to a digital system [11]. This loss of time would almost be eliminated when using MTTWeb.

When asked how they feel about the MTTForms compared to their current system, most teachers found that they were harder to use and that it would take them longer to enter data with the DPPT than normally. This might be due to the fact that the teachers had experience and at least a full day of training with their respective systems whereas the MTT system was relatively new to them. On the other hand, the majority of the teachers claimed that MTTWeb gives them a better overview over how their students perform than they have now. This is unexpected since the low number of students taught by each teacher should give them good personal contacts to every student. If a school with ten students per class can already benefit from the additional information presented by MTTWeb, normal German schools should gain much more.

Generally the feedback for the web interface was much better than that for the DPPT. All interviewed teachers could imagine both using MTTWeb regularly for their class administration tasks and using it at home. The general concept of the system seems to be quite understandable for the teachers and many of them added that the idea had high potential and produced interesting results. The only negative feedback collected here that was not directly connected to technical issues of the prototype was that the system only makes sense in larger schools. One of the reasons for this high acceptance of the web based system is probably that the teachers are already accustomed to using a digital system for their administrative tasks. 75% of the interviewed teachers said that they already used a digital system such as Microsoft Excel® to handle grading. The attendance tracking is fully digital in the ISB now

---

[9] One outlying measurement was not included in the calculations.

anyway, as mentioned above. It is likely that teachers in other schools are less open to the idea of using a computer based system, as indicated by a few informal interviews we did with German teachers.

14.4 Conclusion

Digital pen and paper technology is very promising with regards to classroom administration but has quite a few more hurdles to overcome before it can actually be used. Our testing as well as several other studies showed that the reliability of the collected data is far too low to use it unsupervised. This is not only due to the lacking character recognition but also due to system inherent and technological problems. Advantages in the intelligent interpretation of the strokes collected by the pen could get the technology a lot closer to being useful. More sophisticated software could take context into consideration and for example recognize a stroke that starts inside a box as belonging fully into it, even if the user crosses the border.

A certain understanding of the underlying technology is still needed to avoid mistakes and therefore disproves our assumption that one can simply replace normal pens with digital ones and continue original workflows. At the current state of the technology, extensive training of the users is required, even if they have a considerable understanding of computer technology. Interestingly, teachers had more issues with the DPPT than with the web interface – opposing the initial assumption. That said, the acceptance of the technology in both our study and the results from other scientists has been very good and the end users saw the general idea as a viable solution to some of their problems. Further studies have to be done to find out if DPPT in this form can actually save teachers time, but the prospects look good. Digital pen and paper technology remains a promising tool that might be made viable through further advancements in interpreting technology.

Acknowledgments We would like to thank our supervisors Prof. Dr. Andreas Breiter and Dipl.-Inf. Angelina Lange at the University of Bremen for making this project possible through their support and funding and the members of the MTT project for their cooperation. Additional thanks go to the teachers at the International School of Bremen and their principal whose support allowed us to test the system under real life conditions and the bendit GmbH for providing us with the technical infrastructure we needed.

References

1. C. Bode, T. Mülchen, and B. Ronnenberg, A Framework for Early Detection and Management of School Absenteeism by Using Digital Pen and Paper Technology for Attendance Tracking in German Schools, unpublished.
2. R. Bogdan, S.J. Taylor, Introduction to qualitative research methods, New York: Wiley, 1975.
3. K. Brettfeld, D. Enzmann, D. Trunk, P. Wetzels, Das Modellprojekt gegen Schulschwänzen (ProgeSs) in Niedersachsen: Ergebnisse der Evaluation, [Online] September 2005. Available: http://www2.jura.uni-hamburg.de/instkrim/kriminologie/Online_Publikationen/Kurzbericht_ ProgeSs.pdf.

4. F.D. Davis, R.P. Bagozzi, P.R. Warshaw, User Acceptance of Computer Technology: A Comparison of Two Theoretical Models, Management Science, 35(8), August 1989, 982–1003.
5. C. Despont-Gros, et al., The Digital Pen and Paper: Evaluation and Acceptance of a New Data Acquisition Device in Clinical, Methods of Information in Medicine, 2005, 3, 44.
6. P.C. Dykes et al., The Feasibility of Digital Pen and Paper Technology for Vital Sign Data Capture in Acute Care Settings, AMIA Annual Symposium Proceedings, 2006, Washington DC, USA, pp. 229–233.
7. G. Halász, P. Santiago, M. Ekholm, P. Matthews, P. McKenzie, Anwerbung, berufliche Entwicklung und Verbleib von qualifizierten Lehrerinnen und Lehrern. Länderbericht: Deutschland Paris: Organisation für wirtschaftliche Zusammenarbeit und Entwicklung (OECD), 2004.
8. J. Hartig, E. Klieme (Ed.), Möglichkeiten und Voraussetzungen technologiebasierter Kompetenzdiagnostik. Bundesministerium für Bildung und Forschung (Ed.). Bildungsforschung Band 20. Bonn, Berlin, 2007.
9. A. Lange, A. Breiter, Ansätze zur formativen Evaluation des Einsatzes dezenter mobiler Endgeräte zur Administration in Schulen, in Nomadische und "Wearable"-Benutzungsschnittstellen: Entwurfs- und Evaluationsprinzipien für zukünftige Anwendungen, Holger Kenn and Hendrik Witt, Ed. Bauhaus-Universität Weimar, 2007, pp. 39–41.
10. S. Oviatt, A. Arthur, J. Cohen, Quiet Interfaces That Help Students Think, Proceedings of the 19th Annual ACM Symposium on User Interface Software and Technology, 2006, Montreux, Switzerland, pp. 191–200.
11. K.M. Rahamatulla, C. Trappe, Feasibility of Digital Pen and Paper Technology as Data Acquisition Tool in International Schools, Proceedings of World Conference on Educational Multimedia, Hypermedia and Telecommunications, 2008, pp. 832–839, Chesapeake, VA: AACE.
12. J. Wishart, A. McFarlane, A. Ramsden, Using Personal Digital Assistants (PDAs) with Internet Access to Support Initial Teacher Training in the UK, Proceedings of mLearn 2005 [Online]. Available: http://www.mlearn.org.za/CD/papers/Wishart.pdf.

Chapter 15
A Conceptual Model for a Network-Based Assessment Security System
The Virtual Invigilator

Nathan Percival, Jennifer Percival, and Clemens Martin

Abstract The use of computer to assess learning is increasing at colleges and university as the use of technology on campuses increase. The challenge for the instructors at these institutions is to find a way to ensure the integrity of the assessments while still allow students to access network resources during the assessment. A variety of approaches exist that attempt to create an electronic environment that allows students to access only the resources that are permitted. Unfortunately it is nearly impossible to build a system that allow access to the set of resources that a instructor chooses while guaranteeing that no other resources is being accessed. This paper provides an alternate approach to the challenge of securing an assessment and presents a model of a system that can be used to ensure the integrity of the assessment even when unrestricted access to the network is provided.

Keywords Technology-enhanced learning · Security · Online assessment · Network-based security

15.1 Introduction

In response to employers' desires for technology literate graduates, technology-enhanced teaching is being implemented in a greater number of campuses and programs across the globe [1]. Students today are part of the "millennial generation" who have always had ubiquitous network access and portable communications devices such as cell phones, PDAs, and iPods. As theses students entered higher education institutions, the use of technology as a tool for learning has increased. Most

N. Percival (✉)
Faculty of Engineering and Applied Science, University of Ontario Institute of Technology, 2000 Simcoe St. N., Oshawa, Ontario, Canada, L1H 7K4
e-mail: nathan.percival@uoit.ca

J. Percival and C. Martin
Faculty of Business and Information Technology, University of Ontario Institute of Technology, 2000 Simcoe St. N., Oshawa, Ontario, Canada, L1H 7K4
e-mail: jennifer.percival@uoit.ca; clemens.martin@uoit.ca

S.-I. Ao et al. (eds.), *Advances in Machine Learning and Data Analysis*,
Lecture Notes in Electrical Engineering 48, DOI 10.1007/978-90-481-3177-8_15,
© Springer Science+Business Media B.V. 2010

schools now expect that students will communicate via email, use word processors, and will desire technology connectivity on campus.

Computers and the internet are continuing to become an integral part of life a university campuses [2]. This increased prevalence has lead to an increase in the use of technology for assignments and laboratory situations [1]. The increased integration of the technology into courses is creating a desire and a need to assess the learning outcomes using those same technologies. Simulation systems are also being used more to allow students to learn various concepts and these system can be used during assessments if a method is available to ensure students only use the allowed resources. There is also a significant increase in the use of specialized software for teaching engineering and science courses [3] and this is creating new challenges to the existing methods of securing computers during exams.

According to Ref. [4] as of November 2008 there are over 251 universities and colleges worldwide that have or had some sort of laptop program. These programs range in size from a single program to the entire institution. The number of programs is constantly in flux but generally appears to be increasing.

An informal survey conducted in March 2003 at Brigham Young University found 70% of student that as computers were used more for teaching that they should also be used more for testing [5]. According to Ref. [6], the reliance on paper testing will mischaracterize the accomplishment of students who normally use computers and that this mischaracterization may be significant. The assessment process needs to be reflective of the way the student will need to use the knowledge.

In today's global marketplace, engineering and other disciplines require the skills to analyze an idea during the initial design phase, even before the creation of a prototype, in order to be competitive. To complete this type of analysis, students need to graduate with more hands-on skills with the tools they need to do this type of analysis [7]. Therefore, program that integrate the industry specific technologies into their curriculum and ensure through technology-enhanced assessment that students have mastered the application of theory through the software will provide a great advantage to their students as well as future employers.

Reference [8] suggests that students need to have a comfort with technology and a comfort being tested using that technology. The use of computer based assessment tools that are not the same as the tools used every day is likely to lead to more anxiety and hence decrease the effectiveness of testing and may introduce errors as some students' performances may be reduced by the anxiety. The use of commonly used tools to complete these exams would allow these issues to be reduced [9, 10].

While the ability to test the knowledge of specific software or design and analysis principal using the software is easy for an instructor to design, the ability to do so in an environment that provides a reasonable assurance that the students are not using the computer as a method of cheating is currently impossible. Reference [11] identified the need for the development of a mechanism to deal with cheating during online assessments. Instructors need tools to ensure that testing of students have access to the network are secured to a level similar to traditional paper examinations [12].

This paper outlines a solution that provides the ability to monitor the electronic communications of students during assessments by providing real-time alerts of suspicious events and creating digital records of all communication events. The system can also allow the instructor to determine that suspected cheating was not actually in violation of the rules of the assessment.

The Virtual Invigilator may be used in situations where the computers for the assessment are already installed in the room (such as a computer lab) or the students may be allowed to bring their own laptops. In the case where students are allowed to bring their own laptops, the content of those machines would not be in anyway monitored or checked, creating an open-book examination environment where the student is allowed to use what they bring with them but nothing else.

This paper will examine the existing technologies for securing computer-based exams and discuss the shortcomings of those solutions. The Virtual Invigilator model will then be presented including the overall concept, as well as the basic technical requirements. This paper will then explain why the Virtual Invigilator is a more secure solution that addresses a wider variety of technology-enhanced assessment conditions than the existing solutions. Finally, the paper will present a conceptual model for the of the virtual invigilator system.

15.2   Securing Technology-Enhanced Assessment Environments

There are a number of challenges when attempting to secure examinations requiring a computer with network access. The type of challenges faced varies with the style and content of the test, as well as the specific network resources required. An examination may simply be conducted using multiple choice and short answer questions on a Learning Management System (LMS). The examination may also be much more complex needing the use of software that requires access to a license server, access to a shared files or templates, and access to a system to submit the files that were created during the exam.

The most basic method of securing an online assessment is a password. Instructors setup an online assessment in a LMS and have the LMS require that the student enter a password before they are allowed access to the assessment. This system has proven ineffective. In one case, a password released seconds before an assessment did not prevent ten students not in the room from completing the assessment [11]. Even the more complex restriction of access to certain IP address ranges does not always stop students, as the student may only need to be near the room in which the assessment is occurring and not physically in it.

The systems that are currently used approach the problem of securing an exam by attempting to create an environment that is impossible to cheat in. These systems attempt to prevent students from cheating by making the assessment the only item on the computer they are able to access. Another approach is to physical monitor the student using addition hardware such as cameras and microphones that try to monitor the students' surroundings. Both of these systems require software to be installed on the students' computers and the monitoring system requires that the student have additional hardware installed on the computer. The requirement to install the security application on the computer means that the student taking the assessment is required to have run a specific operating system and is given full access to the code of the product allowing them to reverse engineer the source code.

15.2.1 Secured Testing Environment Solutions

A number of prototype systems have been proposed to create a secure testing environment. Reference [13] proposed a system using a bootable zip disk and Ref. [14] created a method that worked for a CDROM. Both Refs. [13, 14] limit the systems by preventing access to the network and attempting to ensure that the student cannot use anything other than the software provided. Unfortunately, this prevents students from using a large or complex applications or applications that must communicate with a license server. It also does not provide any way to secure exams that are hosted in a LMS. These models could provide security for a specific type of examination on a specific systems, they are not robust enough to handle most assessments.

Currently, the two best-known commercial products that create a secure environment for conducting assessments through a LMS are Securexam Browser from SoftwareSecure [15] and Respondus Lockdown Browser from Respondus [16]. Both of these products attempt to provide a secured web browser that only allows access to assessments provided by a single LMS by attempting to take control of the entire system. Securexam Browser works with only the two most common LMS, WebCT, and Blackboard. Securexam Browser.

Respondus Lockdown Browser has many of the same limitations as Securexam Browser. The system limits student access by blocking access to the task manager, copy and pasting functions, and function keys. In addition, Respondus Lockdown browser needs software to be installed on the server for some LMS. This software allows the LMS to verify that the assessment is being taken in the Respondus Lockdown Browser [17]. This function could also be beaten by customizing another application to respond in a similar manner.

Another method for breaking the security of both of these products is by using a virtual machine. There are more than 15 virtual machines able to run Microsoft Windows listed on Wikipedia [18]. Based on testing of a recent version of Securexam, it was determined that is detects virtual machines from VMWare and Microsoft Virtual machine but not VirtualBox or Virtuozzo. Respondus Lockdown Browser detection of all virtual environments is unlikely. Both of these software base solutions are limited to a total of three different LMS. These commercial products do not support the two main open-source LMS, Moodle, and Sakai [17].

15.2.2   Video Monitoring Solutions

Video monitoring of student actions has been proposed as alternative method for securing computer-based testing. Reference [19] has developed prototype of these type of security system. This type of secured environment is an integrated solution involving both hardware and software components. The system attempts to monitor the actions of the student writing the exam, as well as both the audible and visual environment around the student. SoftwareSecure is currently developing such as system. The system attempts to detect any "abnormal" changes in the environment and then records them for review by the professor at a later time. The problem for this type of system would be determining when the activity is suspicious. The level of sound and motion change in a quiet room, an office cubical, or a coffee house would all be drastically different, making the detection of only suspicious changes quite challenging.

This type of environment is only useful when a student using the system can be placed in an isolated location. It would be virtual useless when used in a room with a number of other students nearby taking exams because the motion and noise nearby students would constant be identified as interesting event. The video monitoring solution requires equipment for every student taking an assessment leading to additional costs. The equipment would also need to be installed prior to the assessment. If the equipment was installed in a lab, it would either need to be installed and removed for an assessment or there would be an increased risk that the equipment could be damaged or stolen leaving the lab short of equipment during an assessment.

15.3   The Virtual Invigilator System

The Virtual Invigilator is a system designed to secure the electronic communications of an assessment environment. It is designed to assist in the proctoring of an assessment in a controlled location such as a classroom with one or more invigilators monitoring the activity within the room. The system assists by monitoring all network traffic from the computers within the classroom. The system is designed to be hardware and software independent with the only technical requirements being that the networking equipment support monitoring of the traffic. Students with any computer running any operating system and any type of application can be monitored using the system.

The Virtual Invigilator is unique in that it does not attempt to create a secured testing environment on an unsecured piece of hardware nor does it attempt to directly monitor the human that completing the exam. The students are allowed to use their local computers in any way they see fit and the monitoring of human actions to real invigilators. The Virtual Invigilator supports the traditional invigilation process by providing the capability to monitor the portion of the assessment environment that cannot be easily monitored by traditional observations methods. The Virtual Invigilator is attempting to extend to assessments that require students to have access to network resources and computer-enhanced assessment environments.

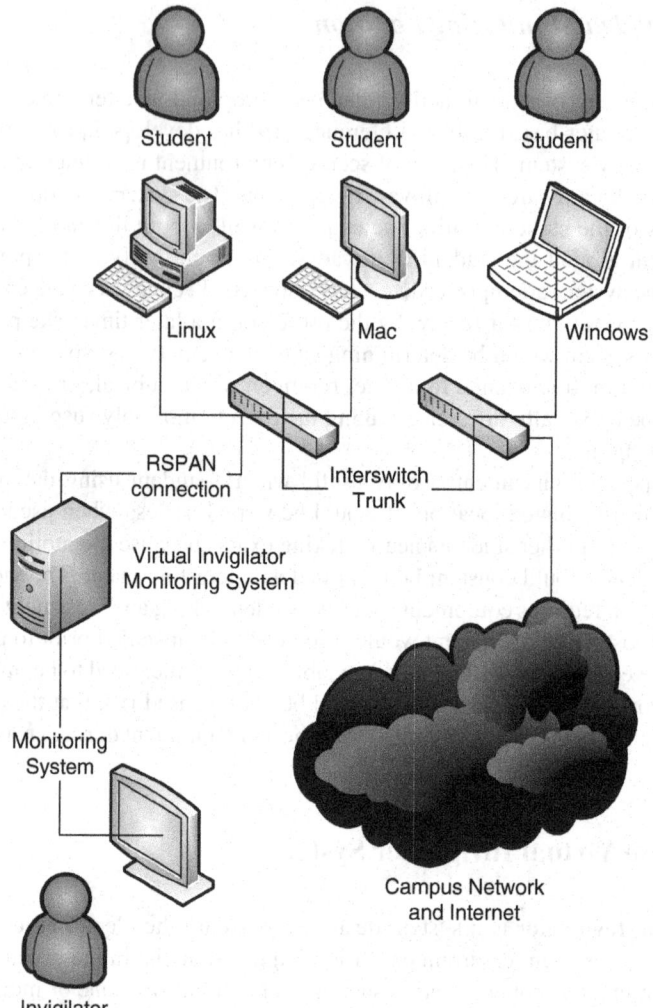

Fig. 15.1 Model system

The overall model of the Virtual Invigilator system shown in Fig. 15.1. The system is designed to operate on a local network. The monitored network only needs to include the network traffic that is either to or from a machine that is in the room being proctored. The network equipment must be capable of providing a copy of all network communications to a single device.

The Virtual Invigilator also access the switches, through Simple Network Management Protocol (SNMP) to identify the physical port to which a computer is connected by identifying the port on the switch that is associated with Ethernet Media Access Control (MAC) address for that computer. The Virtual Invigilator system will need to have a mapping of the physical switch port to actual location in the assessment room to provide to the invigilator if any suspicious activity is found.

The system assumes that the parameters of an assessment can be clearly defined for input into the Virtual Invigilator system. The instructor needs to be able to use a predefined access level (such as only the LMS) or be able to list the sites that the students should be allowed to access. The ability to detect unauthorized events is limited by the specificity of the rules provided.

The Virtual Invigilator is highly customizable and will allow an assessment to be setup through the interface with the criteria by which to detect suspicious activity.

To allow the invigilators to select default setting, the Virtual Invigilator has the ability to be customized at a technical level by staff that are familiar with the requirements of the systems and applications that commonly need to be accessed. This allows the invigilators to have a simple interface for use when preparing a room for an assessment. The customization system can also be used by an instructor to prepare the room for an assessment when the instructor has a complex set of requirements.

15.3.1 Architecture

The network packets are captured by the system and simultaneously processed in two separate ways. First, all traffic is stored to disk for possible later use. Second, the network traffic is processed to identify any violation of the policies that have been setup by the professor in the Virtual Invigilator.

In some networks, there are multiple paths between a sender and a receiver. This means that a packet is not guaranteed to take a specific route from the sender to the receiver. Most networks have a single point, near each network device that all communications with a device must pass. In Ethernet network the network switches have the ability to configure a port on the device to send out a copy of network traffic and not to accept traffic, similar to the way the tap works This is best place to place to tap the network. This may be the only location that can ensure all traffic from a specific computer is captured.

The storage to disk of the data allows the information captured by the Virtual Invigilator to be used to document the events that occurred on the network. Then, if students are caught by the system, even if they claim that the system detected the network traffic in error or that the traffic captured was the result of someone having hacked into their computer this can be fully investigated. Since all network traffic and not only the traffic that was identified as suspicious are recorded, the claims of the student can be easily verified or disproven. The record of the network communications can also be used during the formal appeals process as evidence to support the case against the student who was caught by the Virtual Invigilator.

The second way that the packets are processed is by a real-time event recognition engine. This system will look at individual packets and sets of packets in detail to determine if the packet is acceptable given the rules that have been setup for the assessment. The concept of analyzing data packets as they occur and identifying certain packets or set of packets that are of interest is commonly done on most networks today using Intrusion/Incident Detection Systems (IDS) to find Internet-based security attacks. Most of these systems inspect network packets as they arrive and compare those packets to a set of parameters that help identify suspicious network traffic [20]. IDS system rules currently used are designed to detect network traffic that is malicious in nature such as attempts to hack into a network.

The Virtual Invigilator uses a set of rules that is designed to detect behavior that is contrary to the rules for the assessment that it monitoring. Similar to traditional IDS, the detection process is designed around the assumption that anything that is not explicitly allowed is suspicious traffic and the system should be an alert. In contrast to the software based secure testing system, it is assumed that for most assessments, it is much easier to specify explicitly what is allows rather than to attempt to list every possible thing that is not allowed. The use of new applications or operating systems, new hardware do not have a significant impact of the Virtual Invigilator. Changes to any of these components used to conduct assessment will at most require an adjustment of the rules to handle slight variation in the way these systems operate.

15.3.2 Security Features

To monitor the testing environment, the Virtual Invigilator uses the port-mirroring capability of the network switches to capture a copy of all network packets. All network ports within the room are configured so that a copy of all traffic, in either direction is mirrored to the monitoring port. This ensures that all network packets can be captured, regardless of the configuration of the end machines.

The monitoring port will be connected to the Virtual Invigilator monitoring system. The computer will have this network interface setup in 'promiscuous mode' to allow it to capture all network traffic. When network traffic fails to be classified as acceptable, the system will consider it suspicious and will send a copy of the suspicious packet to the Virtual Invigilator Management system. In addition, the identification of the packet as suspicious will be logged with the recording of the network traffic so that it can be reviewed later. The communication with the management machines is over a separate network from the network being monitor using a second network interface. This allows the monitoring system to communicate with the management system on a network that is not being monitored ensuring the management traffic for Virtual Invigilator systems does not cause alerts.

Once the monitoring system receives a suspicious packet is will provide notification to the invigilator through a visual alert. The monitoring system will analyze the packet and provide as much information as possible to the invigilator about the suspicious network activity. The system will identify to the invigilator the location, based on the network port in use, of the computer creating the suspicious network traffic. The invigilator can then investigate what is happening to make a determination of whether the packet that was flagged suspicious is truly a violation, a false positive, or requires further investigation. The management system will allow the invigilator to flag all suspicious packets as any of these three classifications. The resulting determination by the invigilator will be logged into the monitoring system for analysis after the exam.

After an assessment is completed, the instructor or invigilator can use the management system to thoroughly examine the suspicious activity that was found by the Virtual Invigilator. This allows for the identification of violations not been detected by the original rules. It may also assist the instructor in determining case of collusion that might be hard to find by simply grading a large set of assessments. The false positive results can also be examined by the instructor and by the technical staff to determine if there is some network traffic that could be identified as acceptable and added to the assessment rules allowing more accurate detection in future assessments.

15.3.3   System Design

The Virtual Invigilator system is designed to work with three external entities and the relations are shown in Fig. 15.2. The first of these is the Administration systems that contain data on to authenticate users, determine the physical location of a network port form the switch port and the information to associate a computer with a particular student. A second system is the actual network devices that make up the network. Finally, the network interfaces with the users of the system.

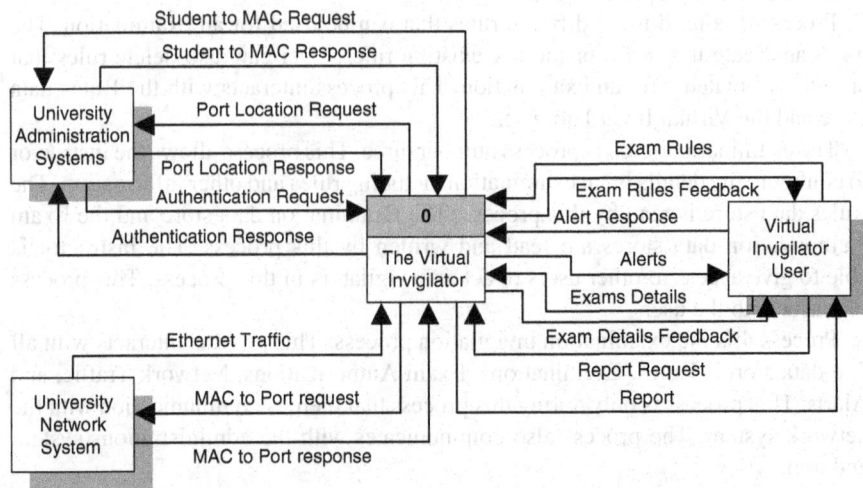

Fig. 15.2 Context diagram

Administrators are able to create rules that can be accessed by all users of the system. They also have the ability to access reports that collect data from various users. Instructors are able to create rules that only they are able to access. They are then able to add use these rules to monitor an examination. The instructor can also allow other users to access the exams they create to invigilating an exam. After an exam is over, the instructor is able to generate reports about exams to determine what infractions occurred during the exam.

The network system provides a copy of all network traffic that either originates from or is destined for a computer with the examination room, including traffic between computers within the examination room. This data is collected by the Virtual Invigilator which stores a copy of the raw data and analyzes whether or not the network traffic is violation of the rules of the examination. The Virtual Invigilator system communicates with the network equipment to determine physical port a computer is connected to.

The university administration system assists the Virtual Invigilator in identifying the who and where of an infraction. The system allows the Virtual Invigilator to convert the information that it receives from the network switch to the physical location in the room. It also allows for the linking of a computer to an individual. A final function of the administration system is to authentication users.

The Virtual Invigilator system has five main processes. The processes, the five data stores that they use, the external entities and the information flows are presented in Fig. 15.3. The processes are based on the actions of the user and their movement through the system.

The first process is the entry point for the Virtual Invigilator user into the system. This process interfaces with the users and the administration system for authorization. It then provides selections of other process and provides authorizations information to the remaining process.

Process 2 is used to modify the rules that can be used for an examination. The user can create new rules or modify existing rules. User can also delete rules that are not associated with an examination. This process interacts with the Rules data store and the Virtual Invigilator user.

The examination setup is process number three. This process allows the instructor to configure the details of an examination including rules and other information. The Rules data store is read by this process. The Examination data store and the Exam Authorization data stores are read and written by this process. The instructor is able to give access to other users to act as invigilators in this process. This process interacts with the user.

Process 4 is the examination invigilation process. This process interacts with all five data stores: Rules, Examinations, Exam Authorizations, Network Traffic, and Alerts. This process is only during this process that there is communication with the network system. The process also communicates with the administrations system and user.

The last process is process five. This process provides reports of various types to the users of the system. In order to do this, the process reads from all five data stores and communicates with the user. The reports are output to the user as requested.

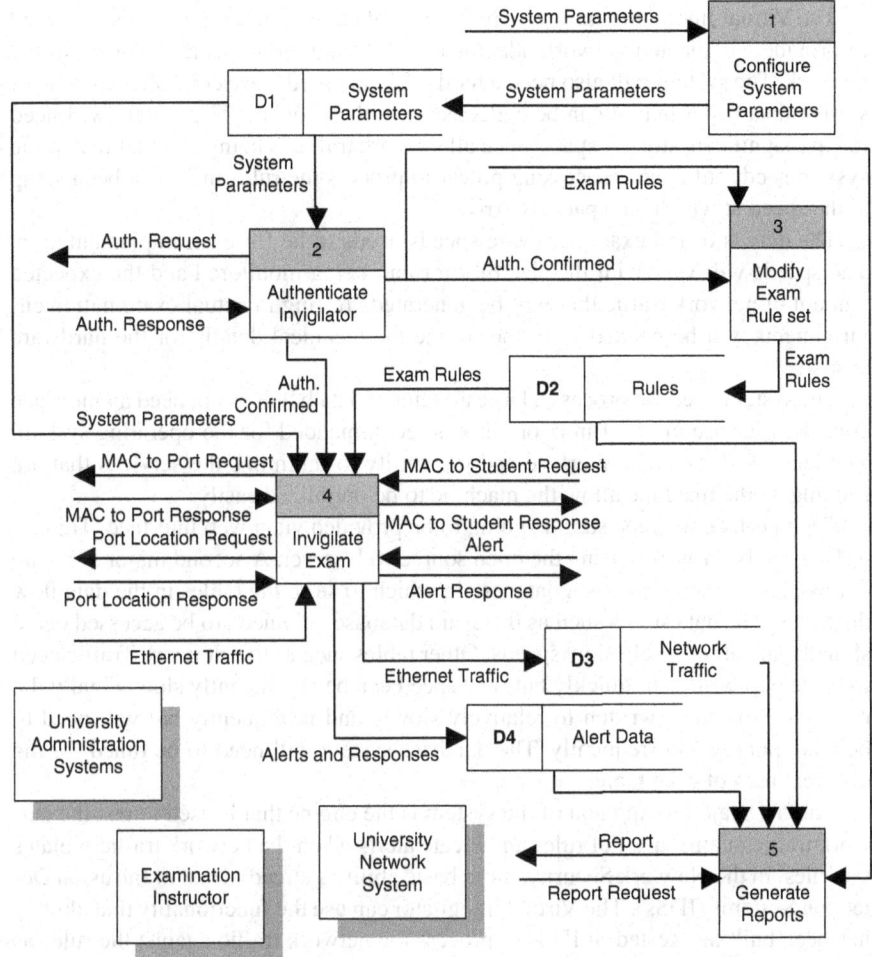

Fig. 15.3 Level 0 data flow diagram

15.3.4 Technical Design – Prototype

The design of the Virtual Invigilator is based on using existing technologies to meet the unique needs of the Virtual Invigilator environment. All components of the system can be built on a single machine. This allows the system to generate minimal network traffic while capturing all the required information.

The user interface for the system will be a web interface. This interface will provide all user interaction. The interface will use Hypertext Transfer Protocol over Secure Socket Layer (https) communication. This will reduce the risk unsecured access to the system and ensure communications with the examination operator is protected from students and others on the network.

The Virtual Invigilator hardware requires dual network interfaces. This is needed to provide a dedicated network adaptor for capturing the network traffic from the network. The system will also need a hard drive and hard drive controller capable of storing data faster than it can be collected from the network. The system will need to have significant storage space since all network traffic is being stored. Finally, the system needs sufficient processing power to process the rules that have been setup at the speed at which data packets arrive.

The details of the exact hardware speeds needs to be for each implementation. The speeds will vary with the size of the room being monitored and the expected amount of network traffic that will be generated. Testing in actual examination environments will be needed to further refine the technical details for the hardware system.

The system needs to process a large amount of data but does not need an interface console. Because of this Linux or Unix is recommended for the operating system. The lack of a graphical interface and the ability to minimize the services that are running on the machine allow the machine to be optimized easily.

The interface to the system is going to be provided via a web interface. This interface can be provided using the open source web server. A second major software component of the system is a database in which to store the tables in the data flow diagrams. The data stores such as the exam database only need to be accessed occasionally and at relatively slow speeds. Other tables such as the Network Traffic need to be able to written to quickly but read speed can be significantly slow. Finally the rules tables need to written to relatively slowly and infrequently but will need to be read quickly and frequently. The database system will need to be tuned for the different uses of each table.

Another major component of the system is the engine that is used to test the network traffic against a set of rules and create alerts when the network traffic violates the rules. In the Network Security, these basic abilities already exist in Intrusion Detection Systems (IDSs). The Virtual Invigilator can use the functionality that already has been built and tested in IDSs to process the network traffic against the rules as setup using the virtual invigilator.

The Virtual Invigilator prototype can be built using open source products. The web interface will be the primary source of control for the users. The user will log into the Virtual invigilator through the web interface and select what they want to do, based on their level of access.

The web interface will access the five data stores housed in the databases. The web interface will have the ability to update most databases but will not only have read access to the network traffic data store. This will help ensure that the data stored in that database is not compromised through the web interface.

15.3.5 Exam Rules

The Virtual Invigilator acts as a tool for instructors to create a set of rules that govern what the students taking an exam are allowed to do. To create these rules there are two starting points: everything is OK and nothing is OK. All exams are based on one of these two starting points. From that, based rules are added to allow or restrict what can be done.

For the starting point of "Everything is OK" all traffic is checked against a set of rules. If the traffic does not match the criteria of any of the rules then it is assumed that the traffic is acceptable. This type of rule checking is most useful during an open book style examination. An instructor can define what the students are not allowed to do and assume that everything else is OK.

The other option for an exam is to assume that no traffic is acceptable except traffic that is explicitly permitted. This type of exam can be used to supervise a closed book exam that needs access to an easily defined set of remote services. With this type of exam, there needs to be rules that allow DNS traffic to flow without generating alarms or every time a name is used instead of an IP address, an alert will occur. The instructor needs to define a list of traffic that can occur without an alert occurring and all other traffic is considered a violation of the exams rules.

For an instructor, it is quite often preferable to use closed-book style exams with a very limited allowable set of systems that can be accessed. This type of rule set will allow the system to capture the largest amount of unauthorized activity but comes with a significantly increased risk of false alerts on traffic. If the instructor has not correctly accounted for and events that occur on computers that users are not even aware of such as Microsoft Windows traffic, software updates and traffic originating from system outside the exam attempting to gain access to the systems within the exam.

For all exams, there is a trade-off with the virtual invigilator of attempting to detect every possible violation versus the rate of false alerts. The use of "everything is OK" based rules reduced the false alerts to only known bad but is likely to have ways that violations can occur and not be detected. In contrast, the "nothing is OK" model will create alerts for every piece of network traffic that is not explicitly allowed which may cause so many alerts that the true violations get lost in the noise. For these reasons it is important that Virtual Invigilator be configured carefully to maximum its usefulness.

15.4 Future Enhancements

A possible enhancement is to provide alerts to invigilators through additional channels. Through the use of portable devices such as Personal Digital Assistants (PDAs), iPhones & Blackberries it would be possible to send alerts to invigilators as they wandered the exam room observing the students. This would allow the invigilator to be investigating an alert and be able to move on the investigating the next alert without needing to return to a computer to access the Virtual Invigilator's web interface.

Finally, functionality can be added to the Report sections by adding tools that will allow an instructor to analyze the traffic from an exam after is occurred and look for violations of the rules that were not originally detected. This could be accomplished by replaying the network traffic through a revised set of filters and providing the new alerts. This would allow instructors to look for events where they believe that a violation occurred that was not detected.

References

1. S. Elwood, C. Changchit and R. Cutshall, Investigating students' perceptions on laptop initiative in higher education: an extension of the technology acceptance model, Campus-Wide Information Systems, vol. 23, pp. 336–349, 2006.
2. D. G. Brown, J. J. Burg and J. L. Dominick, A strategic plan: for ubiquitous laptop computing, Communications of the ACM, vol. 41, pp. 26–35, January 1998.
3. J. Percival and N. Percival, Engineering students' perception of mobile learning, in World Congress on Engineering and Computer Science 2008, pp. 465–470.
4. R. Brown, Colleges with laptop or notebook computer requirements: 2008(Nov 22), p. 1. Available: http://www2.westminster-mo.edu/wc_users/homepages/staff/brownr/NoteBookList. html.
5. S. L. Howell, E-Learning and Paper Testing: Why the Gap? Educase Quarterly, November 4, 2003.
6. M. Russell and W. Haney (2000), Bridging the gap between testing and technology in schools, Education Policy Analysis Archives 8(19), Available: http://epaa.asu.edu/epaa/v8n19.html.
7. H. R. Jacobs. The utilization of a mobile computing environment in undergraduate education. Proceedings of Frontiers in Education Conference, 1996. 26th Annual Conference.
8. L. Baker-Eveleth, D. Eveleth, M. O'Neill and R. W. Stone, Helping student adapt to computer-based encrypted examinations, Educase Quarterly, pp. 41–46, November 3, 2006.
9. G. Sim, P. Holifield and M. Brown, Implementation of computer assisted assessment: lessons from the literature, ALT-J Research in Learning Technology, vol. 12, pp. 215–229, September 2004.
10. L. Baker-Eveleth, D. Eveleth, M. O'Neill and R. W. Stone, Enabling laptop exams using secure software: applying the technology acceptance model, Journal of Information Systems Education, vol. 17, pp. 413–420, 2007.
11. A. B. Campbell and R. P. Pargas, Laptops in the classroom, in Technical Symposium on Computer Science Education: Proceedings of the 34th SIGCSE Technical Symposium on Computer Science Education, 2003, pp. 98–102.
12. L. Guernsey, For Those Who Would Click and Cheat, The New York Times, April 26, 2001.
13. C. C. Ko and C. D. Cheng, Flexible and secure computer-based assessment using a single zip disk, Computers & Education, vol. 50, pp. 915–926, 2008.
14. M. C. Carlisle and L. C. Baird. Design and use of a secure testing environment on untrusted hardware. Presented at Information Assurance and Security Workshop, 2007. IAW '07. IEEE SMC.
15. Software Secure Inc. (2008), Securexam – securexam browser. 2008(6/28/2008), Available: http://www.softwaresecure.com/browser.htm.
16. Respondus Inc. (2008), Respondus lockdown browser. 2008(6/28/2008), Available: http://www.respondus.com/products/lockdown.shtml.
17. Respondus Inc. (2008), FAQs about respondus LockDown browser (LDB). 2008(June 4), Available: http://www.respondus.com/lockdown/faq.shtml.
18. Anonymous (2008, June 17). Comparison of virtual machines. 2008 (June 19), Available: http://en.wikipedia.org/wiki/Comparison_of_virtual_machines.
19. C. C. Kong and C. D. Cheng, Internet examination system based on video monitoring, Internet Research: Electronic Networking Applications and Policy, vol. 14, pp. 48–61, 2004.
20. P. Innella (2001, Nov 16). The evolution of intrusion detection systems. 2008(6/28/2008), Available: http://www.securityfocus.com/infocus/1514.

Chapter 16
Incorrect Weighting of Absolute Performance in Self-Assessment

Scott A. Jeffrey and Brian Cozzarin

Abstract Students spend much of their life in an attempt to assess their aptitude for numerous tasks. For example, they expend a great deal of effort to determine their academic standing given a distribution of grades. This research finds that students use their absolute performance, or percentage correct as a yardstick for their self-assessment, even when relative standing is much more informative. An experiment shows that this reliance on absolute performance for self-evaluation causes a misallocation of time and financial resources. Reasons for this inappropriate responsiveness to absolute performance are explored.

Keywords Engineering education · Self-evaluation · Academic performance · Study plans

16.1 Introduction

Students in engineering programs have limited resources with which to improve their marks. Therefore, it is critically important that they allocate these scarce resources to the areas that have the most room for improvement. This paper focuses on which performance measures (relative or absolute) that students use to self-evaluate as well as how to allocate their study resources. Beyond the classroom, people face opportunities for evaluation. For example, employees receive regular performance evaluations. Both this type of evaluation as well as student grades often serve the purpose of determining someone's future earnings potential via the availability of good jobs and promotions.

S.A. Jeffrey (✉)
Monmouth University, Leon Hess Business School, Department of Management and Marketing, 400 Cedar Ave., West Long Branch, NJ 00764-1898
e-mail: scottajeffrey@gmail.com

B. Cozzarin
University of Waterloo, Department of Management Sciences, 200 University Ave. West, Waterloo, Ontario, N2L 3G1,
e-mail: brian.cozzarin@gmail.com

S.-I. Ao et al. (eds.), *Advances in Machine Learning and Data Analysis*,
Lecture Notes in Electrical Engineering 48, DOI 10.1007/978-90-481-3177-8_16,
© Springer Science+Business Media B.V. 2010

These external evaluations can be based on absolute performance (e.g. meeting a sales or production quota or achieving a given grade on a test), relative performance (e.g. performing in a specific percentile of the workforce or your ranking in a course), or some combination of the two. In most evaluative situations, people will learn of both their relative and absolute performance. For example, in a golf match, a player will know their absolute score and where that score placed them in the final rankings. In the world of professional golf, all that matters is your ranking at the end, even though absolute performance levels are somewhat correlated.

This paper offers reasons and evidence that students focus too much on their absolute performance levels, even in situations where (like in a golf tournament) relative standing is all that matters. In the first section of this paper, we review the literature on self-evaluation and the predictions regarding resource allocation that this literature makes. Then we will explore some reasons why students will rely too much on their absolute performance. We then report a study that shows an over-reliance on absolute performance when evaluating performance satisfaction and the decision to expend resources towards self-improvement. The final section closes with some implications and directions for future research.

16.2   The Psychology of Self-Evaluation

There has been a great deal of psychological research on how people select the criteria and comparators to use when evaluating their own abilities and performance. Beginning with work by Festinger [5], researchers have proposed that people seek out objective information when attempting to evaluate their abilities. In the absence of objective performance information, they compare themselves to similar others, since it was originally hypothesized that the main motivation of social comparison was accurate self-knowledge [5, 13].

As research on social comparison developed, other additional motives were explored. Other research on social comparison has extended this to include two additional motives that would drive the selection of comparators. When people were attempting to self-enhance, they would use downward comparisons [2]. When people had a self-improvement motive, they often would compare upwards, giving them a target or goal [15]. In an academic setting, most students use upwards social comparison in order to motivate themselves to do better [7, 14].

We hypothesize that if students' self-esteem were threatened by low absolute performance, then the natural tendency predicted by social comparison would be to try to find a downward comparison target [4]. This would cause students to ask for distributional information, as is often the case. By seeing that they were above a certain number of students in the class would appease this negative reaction. Note that this would predict a focus on relative performance levels when reporting satisfaction, particularly for low absolute/high relative performers.

Other research on social comparison has shown that students do not choose downwards social comparison but rather choose upwards comparisons provided they feel that they are able to improve their performance [8, 11]. This would be consistent with a prediction that students would use absolute performance levels to evaluate their abilities. It would also support a prediction that students would use absolute performance levels to allocate resources, given the assumption that they have adopted an improvement motive.

A related stream of research hypothesized that people also compare themselves to previous versions of themselves or future ideal versions [1]. If students were using previous versions of themselves, then a low absolute score would be discouraging, especially for students who had consistently received high absolute scores. Therefore, low absolute scores would trigger a self-improvement motive that would cause them to dedicate more effort to areas in which their absolute score was low. This could also trigger the common request students make for distributional information to handle short term disappointment. However, once the self-improvement motive becomes dominant, students will expend effort where their absolute performance is lower, relative performance held constant.

We believe that in circumstances where people receive both relative and absolute performance feedback, people will place too much weight on their absolute performance when it is less important than relative performance. For example, consider grades in university. For students, grades carry nearly as much, if not more, importance than money. This is reasonable since better grades tend to be correlated with higher salaries [12]. Grades are determined by performance on exams, but in courses that are curved on a forced grade distribution, performance relative to others is the more important metric. The same is true of a golf tournament, where relative standing will determine prize earnings as well as competitions for promotions within organizations [10]. The important performance benchmark in all of these examples is being above a certain number of one's rivals.

Continuing with the grading example, suppose a student is taking two classes and has received his or her midterm grade. One course is known for a hard teacher who gives hard exams and grades them strictly. The other course is taught by a far more lenient professor. In the difficult class, the student receives a score of 65 out of 100 while in the easy class he or she scores 85 out of 100. Further suppose that this student receives information that both of these scores place him or her in the 80th percentile of performance. Since both classes are graded on a curve, he or she should be equally satisfied with the performance on both exams; however we predict that our hypothetical student will be much less satisfied with the 65. While much research has shown that people do use relative comparison when evaluating their own performance [6,9,11], we believe it is difficult to ignore absolute performance, even when it is completely irrelevant.

The final reason it is difficult to ignore absolute performance is because often times it actually is a valid signal of performance. In school, we are socialized to learn that absolute performance is a signal of learning, even though there is some arbitrariness in the grading process; some teachers create difficult tests, and sometimes grade assignments more strictly. This fact should make relative class standing is a better indicator of performance. All through school children are taught that 90% is an "A", 80% is a "B", all the way down to 50% or below which was an "F".

The natural reaction to scoring 65 out of 100 is that this is a low C or even a D. This could lead a student to question their knowledge of the material in the course and lower the satisfaction with their performance.

A related explanation for this is that students may feel that a test should be representative of the knowledge a student should have received in class. In other words, getting 65% right could be interpreted that the student only understood 65% of the material. This would also lead to below average self-assessments, especially if the student felt he or she knew the material well. All of these reasons suggest that students will have a hard time ignoring absolute performance, even in situations where it is not important or even irrelevant. We therefore hypothesize that relative performance held constant, students will be more satisfied with higher absolute performance.

While satisfaction with results is important, the behavioral consequences of those feelings are more important. Suppose that instead of equal relative ranking, the higher absolute score put our hypothetical student at a lower relative standing than the score on the more difficult test. If his or her intention to study is based upon her satisfaction level, and satisfaction is inappropriately affected by absolute performance, then he or she will spend too much time studying for the class in which they are in better relative position. It is probably not the case that a student evaluates his performance solely on absolute performance. However, in cases where relative performance is far more important, an over-reliance on absolute performance will cause a misallocation of improvement resources. Therefore, we expect to find that students will devote more resources (time, money, etc.) to improving performance in areas where absolute performance is lower, relative standing held constant.

16.3 Experiment

16.3.1 Method

Sixty engineering students at a large Canadian university were asked to complete a survey. This survey asked participants to imagine that they had just received feedback regarding their performance on a midterm exam (please see Appendix for instrument). Half of the participants were told that they were taking a class from a notoriously difficult professor with a reputation for tough grading standards, while the other participants were told that they were taking a class from a more lenient professor. All participants were told that the professor used a grading curve in which the top 30% of students receive an A; the middle 60% receive a B; with the rest of the students receiving a C or D. So that the participants could better understand the performance feedback, they were told that there were 100 students in the class, so that these percentages could be easily translated into the number of students who would receive each grade. Students in the more difficult section were given lower absolute scores than those in the more lenient professor's class. All students were

Table 16.1 Summary
of conditions

	Strict professor	Easy professor
Grading targets:		
A	30%	30%
B	60%	60%
C or D	10%	10%
Exam results:		
Mean score	55	72
Participant score	52	72
Relative position in class	53rd	53rd

told that their performance on the test placed them in the 53rd percentile. Table 16.1 is a summary of the conditions faced by the survey participants.

All participants were asked to state their satisfaction with their performance. Participants were then told that they were planning how to allocate their time to studying for the final exam in the same course. Specifically, participants were asked how long they believed that they would study for the final exam, using a closed end scale from 2 to 16 h in increments of 1 h. After answering this question, they were informed of an Internet course that would help them prepare for the final and asked their level of interest in the course and how much they would be willing to pay for that course. To avoid problems with modulus mapping (extreme values), they were asked to provide a number between zero and $100. They were told that if they would not take the course regardless of price, they should enter zero. Finally, demographic information was collected from students, to see if there was a gender or age effect on the results.

Because of the cover story regarding the professor, participants should realize that their absolute performance on the midterm is a nearly meaningless signal as to how they are doing in the course and how much time and other resources they should expend in preparing for the final. This would imply that there should be no differences in responses between conditions. However, if participants are sensitive to their absolute performance as predicted, then participants in the difficult course will predict spending more time studying for the exam and be willing to pay more money for the Internet preparation course.

Across the entire sample, students seemed neither satisfied nor dissatisfied with their performance. They were willing to spend approximately 8 h studying for the final exam, spend approximately $20.00 on test preparation, and 68% of them stated a positive (i.e. non zero) willingness to pay for the course. There were differences across conditions however, as will be discussed in the following section.

16.3.2 Results

We conducted tests to ascertain whether the groups were similar to each other in terms of the demographic variables. In the low absolute score condition there were 8 women and 21 men, while in the high absolute condition there were 8 women and

Table 16.2 Means by conditions

	Low absolute	High absolute	Difference
Satisfaction	2.69	3.58	0.89**
Intent to study (h)	9.03	7.35	1.68†
Willingness to pay	$26.98	$13.29	$13.69*

†p < .10, *p < .05, **p < .01 in two-tailed t-test.

Table 16.3 Regression results

	Dependent variables			
	Satisfaction	Study plans	WTP	Positive WTP
Age	0.01	0.68*	0.77	0.11
Gender	0.24	−0.62	−3.53	0.09
Absolute score	0.84**	−1.83	−15.24*	−1.70*
Satisfaction	NA	0.07	2.48	0.001
Fit statistics	F(3,55) = 2.52	F(4,54) = 2.55	F(4,54) = 1.61	
	R-square = 0.12	R-square = 0.16	R-square = 0.11	R-square = 0.18

†p < .10, *p < .05, **p < .01.

22 mean, $F(1,57) = 0.006$, ns. Average age in the low absolute condition was 20.83 and 20.73 in the high absolute condition, $t(57) = 0.23$, ns. Since the make up of each group was almost identical, we feel comfortable in comparing results across conditions.

As can be seen in Table 16.2, subjects were more satisfied with higher levels of performance when their absolute score was higher, $t(58) = 2.87$, p < .01. With respect to the willingness to allocate resources, participants expressed a higher willingness to pay when their absolute performance was lower, $t(58) = 2.39$, p < .05. Their intention to study was also higher (although only marginally so), two-tailed $t(58) = 1.80$, p = .08.

A more thorough analysis was provided by an OLS regression, which also included the demographic information of age and gender. These results can be seen in Table 16.3. The addition of demographic information does not change the result that satisfaction is positively impacted by absolute score, with relative position held constant, $\beta = 0.84$, $t(55) = 2.663$, p < .01. Since satisfaction with performance might also predict the willingness to allocate resources we included that in the two regressions (willingness to pay and study plans). The coefficients on this variable did not approach significance for either resource allocation question. In addition, it did not cause the significance of the level of absolute score to change. This rules out the change in resource allocation being mediated by satisfaction [3].

A second means by which to test the willingness to take an Internet preparation course is by checking the proportion of students in each condition that would not take the course regardless of price. To analyze this question, a binary variable was created such that any participant responding zero dollars to the willingness to pay question was coded as someone who would not enroll in the course (i.e. Positive WTP = 0). Those that entered any positive number were coded as someone who would enroll in the course (i.e. Positive WTP = 1). The proportion of participants reporting a positive willingness to pay was 86.2% when absolute performance was low, and only 51.6% when it was high, $t(58) = 2.88$, $p < .01$. This difference was confirmed in a logistic regression also reported in Table 16.3. None of the control variables predicted the willingness to take the course, nor did the overall satisfaction level with the performance.

Age was the only demographic variable that explained variance in any of the dependent variables. Specifically, older students reported a higher willingness to study with an extra 68 h per every year of age, this is most likely explained by the fact that at Canadian Universities, only third and fourth year marks are counted towards graduate school entrance. Age had no effect on satisfaction with performance or willingness to pay for an internet preparation course. Gender was insignificant across all independent variables, with no differences in satisfaction, willingness to study, or willingness to pay.

Absolute score was analyzed as a dummy variable, with a value of one for the high absolute score (72) and zero for the low absolute score (52). This makes the effects easy to calculate. As can be seen from the regression results, the level of absolute score had the predicted effects on all dependent variables. Students receiving a higher absolute score were nearly a full point more satisfied with their performance (on a six-point scale). Those receiving a lower absolute score reported a willingness to study for two more hours. Lower performing students (in absolute terms) were also willing to pay approximately $15 more for the internet preparation course. They were also much less willing to report a zero willingness to pay as shown by the negative coefficient in the logistic regression.

It is interesting to note that while satisfaction levels were effected by absolute performance, satisfaction levels did not affect intentions to study or willingness to pay for prep courses. This suggests that students are taking their cues from performance more than simply from how they feel about their performance.

16.3.3   Discussion and Conclusions

This study showed that engineering students are more satisfied with higher absolute levels of performance. This in and of itself is not surprising, as people should be more satisfied with better performance. The real contribution here is that resource allocation follows from absolute performance levels even when relative standing, all that matters to their performance, is held constant. Students in our sample stated a willingness to spend more time studying and more money for preparation courses when their absolute performance was lower, even when their relative position was all that mattered.

We believe that this is a mistake, with long term negative consequences. Resources should be allocated to the item that will do the most long term good, and that should be to activities that will increase marks, not necessarily absolute performance levels. If studying resources are misallocated, then overall grades will suffer. While our study held relative performance constant, imagine a real situation where a student actually has lower relative standing in a course where he or she receives a high absolute grade. In this case, it is a clear mistake to allocate resources away from that course towards a course where he or she ranks higher but scored lower in absolute terms.

From a professor's standpoint, if a professor wants students to spend more time studying the material in their course, then they should be strict graders on assignments during the term. The problem is that strict graders tend to receive poor teaching evaluations. Therefore, there is a mixed incentive for faculty members. In order for their students to take their topic seriously, they should grade strictly, yet their own performance evaluations may suffer, affecting their salary as well as tenure and promotion decisions.

In the end, it is in the professor's best interest to try to effectively communicate the grading criteria, and be certain that students understand exactly where they stand with respect to their overall final performance. While this seems easier said than done, to be true educators, professors must try to avoid inflating students' marks so that they can increase their own teaching evaluations. In the end, the best strategy would be to create assignments and tests that reflect true understanding of the material so that students "curve themselves". This would then make absolute performance proxy nicely for relative standing.

Limitations and Future Research. One limitation of this study is that it is hypothetical, asking students to self-report their intention. The results would be more compelling with a realistic study where real grades were at stake. The authors don't believe that this would be ethical (and are sure the ethics group would agree). However, the authors believe it is reasonable to assume that students have adequate psychological access to their future behaviors with respect to studying behavior. Also, there is little reason to believe that students would not respond accurately to these questions. Perhaps future research can tie incentives to overall performance levels to see if this generalizes to actual behavior.

Another potential limitation to this study is the binary nature of the absolute versus relative performance level. It would be interesting to study the effect of grading schemes that depend on mixed levels of relative and absolute performance. While the authors believe that students would continue to overweight absolute performance, it would be beneficial to understand how responsive students are to changes in the weighting of relative performance.

Finally, there is a question as to whether these results will generalize beyond a group of Canadian engineering students. Fundamentally, we have little reason to believe that students elsewhere in the world would view absolute performance levels any differently. One potential issue is the move in many universities to a numeric grading system and away from the letter grades historically given in many North American universities. To the extent that percentage grading makes "curving" difficult, these results would be less appropriate. However, even under a strict percentage system, additional marks can be reserved by faculty members for final adjustments, thus grading on a curve is still possible in that scenario.

16.4   Conclusion

In most situations, students receive both types of feedback – relative and absolute performance, and must make judgments about how to combine this feedback. The results of this study suggest that students will put too much emphasis on their absolute performance, and underweight their relative standing in a course. This could have negative consequences such as the misallocation of study resources (time and money) targeted at the wrong areas. This misallocation could have negative consequences on overall academic performance.

References

1. Albert, S. (1977). "Temporal comparison theory." Psychological Review **84**(6): 485–503.
2. Aspinwall, L. G. and S. E. Taylor (1993). "Effects of social-comparison direction, threat, and self-esteem on affect, self-evaluation, and expected success." Journal of Personality and Social Psychology **64**(5): 708–722.
3. Baron, R. M. and D. A. Kenny (1986). "The moderator-mediator variable distinction in social psychological research: Conceptual, strategic, and statistical considerations." Journal of Personality and Social Psychology **51**(6): 1173–1182.
4. Buunk, B. P., S. E. Taylor, et al. (1990). "The affective consequences of social-comparison – either direction has its ups and downs." Journal of Personality and Social Psychology **59**(6): 1238–1249.
5. Festinger, L. (1954). "A theory of social comparison processes." Human Relations **7**: 117–140.
6. Festinger, L. (1958). The motivating effect of cognitive dissonance. Assessment of human motives. G. Lindzey. New York, Holt, Rinehart, & Winston.
7. Foddy, M. and I. Crundall (1993). "A field study of social comparison processes in ability evaluation." British Journal of Social Psychology **32**: 287–305.
8. Gibbons, F. X., H. Blanton, et al. (2000). "Does social comparison make a difference? Optimism as a moderator of the relation between comparison level and academic performance." Personality and Social Psychology Bulletin **26**(5): 637–648.
9. Lane, D. J. and F. X. Gibbons (2007). "Social comparison and satisfaction: Students' reactions after exam feedback predict future academic performance." Journal of Applied Social Psychology **37**(6): 1363–1384.
10. Lazear, E. P. and S. Rosen (1981). "Rank-order tournaments as optimum labor contracts." The Journal of Political Economy **89**(5): 841.
11. Michinov, N. and J. M. Monteil (1997). "Upward or downward comparison after failure. The role of diagnostic information." Social Behavior and Personality **25**(4): 389–398.
12. Roth, P. L. and R. L. Clarke (1998). "Meta-analyzing the relation between grades and salary." Journal of Vocational Behavior **53**(3): 386–400.
13. Sedikides, C. and M. J. Strube (1997). Self evaluation: To thine own self be good, to thine own self be sure, to thine own self be true, and to thine own self be better. Advances in experimental social psychology. M. P. Zanna. San Diego, CA: Academic Press, vol. 29, pp. 209–269.
14. Wayment, H. A. and S. E. Taylor (1995). "Self-evaluation processes: Motives, information use, and self-esteem." Journal of Personality **63**(4): 729–757.
15. Ybema, J. F. and B. P. Buunk (1993). "Aiming at the Top – Upward Social-Comparison of Abilities After Failure." European Journal of Social Psychology **23**(6): 627–645.